Edexcel

foundation

GCSE Mathematics

practice book

Keith Pledger

Gareth Cole

Peter Jolly

Graham Newman

Joe Petran

www.heinemann.co.uk
✓ Free online support
✓ Useful weblinks
✓ 24 hour online ordering

01865 888058

Heinemann

Inspiring generations

Heinemann Educational Publishers
Halley Court, Jordan Hill, Oxford OX2 8EJ
Part of Harcourt Education Limited

Heinemann is the registered trademark of
Harcourt Education Limited

© Harcourt Education Ltd, 2006

First published 2006

10 09 08 07
10 9 8 7 6 5 4 3 2

British Library Cataloguing in Publication Data is available from the British Library on request.

10-digit ISBN: 0 435 53362 2
13-digit ISBN: 978 0 435533 62 5

10-digit ISBN (10-pack): 0 435 53377 0
13-digit ISBN (10-pack): 978 0 435533 77 9

Typeset by Tech-Set Ltd, Gateshead, Tyne and Wear
Original illustrations © Harcourt Education Limited, 2006
Illustrated by Theresa Tibbetts
Cover design by mccdesign
Printed in China by CTPS
Cover photo: PhotoLibrary.com

Acknowledgements
Harcourt Education Ltd would like to thank those schools who helped in the development and trialling of this course.

This high quality material is endorsed by Edexcel and has been through a rigorous quality assurance programme to ensure that it is a suitable companion to the specification for both learners and teachers. This does not mean that its contents will be used verbatim when setting examinations nor is it to be read as being the official specification – a copy of which is available at www.edexcel.org.uk

The publisher's and authors' thanks are due to Edexcel Limited for permission to reproduce questions from past examination papers. These are marked with an [E]. The answers have been provided by the authors and are not the responsibility of Edexcel Limited.

The authors and publisher would like to thank the following individuals and organisations for permission to reproduce photographs:

Science Photo Library p1; photos.com pp4, 7; Corbis p66; Alamy Images p124; Empics p153

Every effort has been made to contact copyright holders of material reproduced in this book. Any omissions will be rectified in subsequent printings if notice is given to the publishers.

Tel: 01865 888058   www.heinemann.co.uk

Contents

About this book

This revised and updated edition provides a substantial number of additional exercises to complement those in the Edexcel GCSE Mathematics foundation textbook. Extra exercises are included for almost every topic in the course textbook.

The author team is made up of Senior Examiners, a Chair of Examiners and Senior Moderators, all experienced teachers with an excellent understanding of the Edexcel specification.

Clear links to the course textbook help you plan your use of the book.

Please note that the answers to the questions are provided in a separate booklet. This will be supplied with each order for 10-packs, or with your first order for single books.

1 Understanding whole numbers

1 Draw a place value diagram and write in

(a) the number 5348

(b) a four digit number with a 7 in the thousands column

(c) a three digit number with a 4 in the hundreds column

(d) a two digit number with a 5 in the tens column

(e) a six digit number with a 4 in each column

(f) a three digit number with a 4 in the first and last columns

(g) a four digit number with a 9 in the first and last columns

(h) a four digit number with a 7 in the units column and digits less than 7 in all other columns.

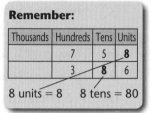

Remember:

Thousands	Hundreds	Tens	Units
	7	5	**8**
	3	**8**	6

8 units = 8   8 tens = 80

2 Write these numbers in words.

(a) 47   (b) 73   (c) 128   (d) 207

(e) 1300   (f) 4236   (g) 32 406

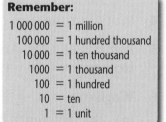

Remember:

1 000 000 = 1 million
100 000 = 1 hundred thousand
10 000 = 1 ten thousand
1000 = 1 thousand
100 = 1 hundred
10 = ten
1 = 1 unit

3 Use the place value diagram to help you write these numbers in words.

(a) 4563   (b) 64 500   (c) 302 709

(d) 6 538 741   (e) 13 773 406   (f) 45 000 000

(g) 23 500 004

4 (a) The distance from the Earth to the Moon is 400 000 kilometres. Write this number in words.

(b) The distance from the Earth to the Sun is ninety-three million miles. Write this number in figures.

5 The attendance figures at six major sporting events are listed below. Write the numbers in order of size, starting with the smallest.

Boxing match: five thousand
Soccer match: 42 000
Hockey international: seven thousand

Horse race: 120 000
Snooker: 850
Rugby cup final: 80 000

6 Put the numbers in the cloud in order of size. Start with the smallest.

17   71
710
107   170

Exercise 1.2 **Links 1C–E**

1 You can use the number line to help you answer these questions.

   (a) Increase 7 by 5   **(b)** Decrease 13 by 4

   (c) Decrease 5 by 5   **(d)** Increase 2 by 17

   (e) Decrease 20 by 3   **(f)** Increase 14 by 6

   (g) How much is the increase from 7 to 15?

   (h) How much is the decrease from 16 to 5?

   (i) What change moves 8 to 14?

   (j) What change moves 13 to 2?

   (k) What change moves 19 to 1?

2 Work out

   (a) the total of 17 and 46

   (b) 68 plus 32   **(c)** 427 + 86

3 A coach has 54 seats downstairs and 38 seats upstairs. How many seats does the coach have altogether?

4 Work out the sum of all the whole numbers from 10 to 20.

5 Ben has 48 CDs. Angela has 25 CDs. How many CDs more than Angela does Ben have?

6 Work out

   (a) the difference between 200 and 86

   (b) 3000 − 238

   (c) the number which is 17 less than 1001

   (d) 320 minus 18

7 When it was new the value of a car was £16 580. When it was five years old the value of this car was £4600. Work out the difference between the value of the car when it was new and the value of the car when it was five years old.

8 Paddy will have his 45th birthday in the year 2037. In which year was Paddy born?

Number line (right side):
20, 19, 18, 17, 16, 15, 14, 13, 12, 11, 10, 9, 8, 7, 6, 5, 4, 3, 2, 1, 0

Add (+)	**Subtract (−)**	**Multiply (×)**	**Divide (÷)**
These words mean you have to **add**:	These words mean you have to **subtract**:	These words mean you have to **multiply**:	These words mean you have to **divide**:
Increase Total Sum Altogether	Decrease Difference Minus Take away	Product Times	Goes into Share Quotient

Exercise 1.3

1 Multiply these numbers by **(i)** 10   **(ii)** 100   **(iii)** 1000
  (a) 7   **(b)** 16   **(c)** 243
  (d) 547   **(e)** 400   **(f)** 801

> **Remember** the different ways of multiplying: traditional method, grid method and Napier's bones.

2 Work out
  (a) 23×5 **(b)** 32×7 **(c)** 54×6
  (d) 320×4 **(e)** 527×8 **(f)** 340×6
  (g) 1984×2 **(h)** 3265×3 **(i)** 999×9

3 Work out
  (a) 23×50 **(b)** 32×700 **(c)** 34×60
  (d) 632×200 **(e)** 460×3000 **(f)** 500×212

4 Divide these numbers by **(i)** 10   **(ii)** 100   **(iii)** 1000
  (a) 3000   **(b)** 54 000   **(c)** 753 000
  (d) 50 000   **(e)** 5 000 000   **(f)** 1 333 000

> **Remember** the different ways of dividing: long division, short division and chunking.

5 Work out
  (a) $36 \div 2$ **(b)** $63 \div 3$ **(c)** $84 \div 4$
  (d) $75 \div 5$ **(e)** $120 \div 6$ **(f)** $256 \div 8$
  (g) $364 \div 7$ **(h)** $483 \div 3$ **(i)** $512 \div 4$

> You should practise doing these **without** a calculator.

6 Work out
  (a) $360 \div 20$ **(b)** $6300 \div 300$ **(c)** $8400 \div 40$
  (d) $120 \div 40$ **(e)** $120\,000 \div 600$ **(f)** $1200 \div 40$
  (g) $300 \div 50$ **(h)** $3600 \div 30$ **(i)** $4800 \div 200$
  (j) $450\,000 \div 9000$ **(k)** $65\,000 \div 2500$

7 Work out
  (a) 345×12 **(b)** 243×34 **(c)** 356×45
  (d) 360×64 **(e)** 283×25 **(f)** 576×34
  (g) 579×87 **(h)** 936×56 **(i)** 1234×56

8 Work out
  (a) $169 \div 13$ **(b)** $196 \div 14$ **(c)** $225 \div 15$
  (d) $441 \div 21$ **(e)** $456 \div 24$ **(f)** $357 \div 17$
  (g) $2160 \div 36$ **(h)** $5280 \div 24$ **(i)** $7320 \div 15$

9 Jennifer's geography coursework is 32 pages long.
  She needs to make a photocopy. The cost of making a
  photocopy is 4p for each page.
  Work out the total cost of making the photocopy of all
  32 pages.

10 Work out

(a) $325 \div 25$ (b) 800 divided by 40

(c) $\dfrac{720}{18}$ (d) the number of 20p coins in £6.

11 Mr Khan won £720 in the Lottery. He shared this money equally between himself, his wife and his two children. Work out how much each person received.

12 Martin has 360 bottles of wine to put into crates. Each crate holds 12 bottles. How many crates will Martin need to use?

Exercise 1.4 Link 1J

1 (a) Lawns Cricket Club scored 205 runs in their first innings and 175 runs in their second innings. How many runs did they score altogether?

(b) In the same match, Nayland Cricket Club scored a total of 345 runs. The team that wins the match is the team that scores the most runs.

(i) Which team won the match?

(ii) By how many runs did they win?

2 Summer earns £6 for each hour she works. Last week she earned £144. How many hours did Summer work last week?

3 Steven was born on 31st October 1981. Work out the date of his 30th birthday.

4 A single entered the charts at number 16. The following week it rose to number 9.

(a) By how places did the single rise in this week?

(b) The next week the single fell 4 places. At what position in the charts was the single then?

5 Megan bought 5 books and 3 pens. She paid £6.50 for each book and £1.05 for each pen. Work out the total amount Megan paid for the 5 books and 3 pens.

6 Sharon bought a box of chocolates. The box contained 24 chocolates. Sharon ate 6 of the chocolates herself. She then shared the rest equally between her 3 cousins. How many chocolates did each cousin receive?

7 An empty school bus picks up 23 students at the first stop, 15 students at the second stop and 9 students at the third stop. It then travels directly to school without stopping. How many students were on the bus when it arrived at school?

8 Tom played an arcade game and scored 19 750. His friend Kenny then scored 20 500. How many more points did Kenny get than Tom?

9 Gina pays 25p for each download of a ringtone for her mobile. One month she downloaded 15 ringtones. How many pence did this cost Gina?

10 Paul delivers milk to his customers. He delivers 500 bottles each day. He carries them all in crates on his milk float. Each crate holds 24 bottles of milk. How many crates does Paul need to hold all 500 bottles of milk?

Exercise 1.5

Links 1K–L

1 Round these numbers to the nearest ten.
   (a) 68   (b) 62   (c) 45   (d) 34
   (e) 186   (f) 245   (g) 312   (h) 32
   (i) 2341   (j) 4788   (k) 666   (l) 8

2 Round these numbers to the nearest hundred.
   (a) 225   (b) 78   (c) 202   (d) 455
   (e) 3148   (f) 2155   (g) 6145   (h) 4670

3 Round these numbers to the nearest thousand.
   (a) 3260   (b) 5780   (c) 3500   (d) 2499
   (e) 10 250   (f) 25 575   (g) 675   (h) 5555

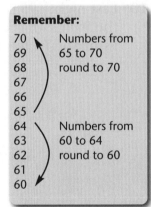

Remember:

70	Numbers from
69	65 to 70
68	round to 70
67	
66	
65	
64	Numbers from
63	60 to 64
62	round to 60
61	
60	

4 Round each of these numbers:
   (a) 364 to the nearest ten
   (b) 4872 to the nearest thousand
   (c) 25 638 to the nearest hundred
   (d) 324 561 to the nearest ten.

5 Last Monday 43 617 commuters arrived at Liverpool Street Station. Round this number to
   (a) the nearest 10
   (b) the nearest 100
   (c) the nearest 1000.

6 Write each of these numbers to 1 significant figure.

    (a) 14   **(b)** 25   **(c)** 37   **(d)** 146

    (e) 231   **(f)** 407   **(g)** 4870   **(h)** 3714

    (i) 1758   **(j)** 6531   **(k)** 106 317   **(l)** 22 493

> When rounding to 1 significant figure (1 s.f.), look at the place value of the first digit.

7 The maximum speed of a train is 142 mph.
Write this number to 1 significant figure.

8 The distance Jane travelled was 1865 miles.
Write this number to 1 significant figure.

9 Showing all your rounding, make an estimate of the answer to these calculations.

> You can check your answer with a calculator.

    (a) $\dfrac{62 \times 49}{32}$ **(b)** $\dfrac{317 \times 402}{198}$ **(c)** $\dfrac{3008 \times 21}{17 \times 43}$

    (d) $\dfrac{38 \times 478}{500}$ **(e)** $\dfrac{567 \times 29}{19 \times 87}$ **(f)** $\dfrac{2780 \times 599}{3250}$

    (g) $\dfrac{666 \times 450}{35 \times 699}$ **(h)** $\dfrac{325 \times 787}{9999}$ **(i)** $\dfrac{414 \times 265}{19 \times 301}$

10 Last night 695 people paid £2.95 each to see a 'Battle of the Bands' in the school's sports hall.
Estimate the total amount of money paid.

2 Number facts

1 Copy the number line and fill in the missing numbers.

−5 __ __ −2 __ 0 1 2 3 4 5

2 On New Year's Day the temperature, in °C, in five cities was

London	Glasgow	Moscow	Athens	Madrid
0	−2	−14	12	8

Write these temperatures in order, starting with the lowest.

3 Find the missing two numbers in each of the sequences.

  (a) 12, 7, 2, __, __, −13

  (b) −16, −12, −8, __, __, 4, 8

4 Find the number which is

(a) 7 less than 4	**(b)** 4 less than 0
(c) 8 more than −3	**(d)** 10 more than −10
(e) 1 less than −5	**(f)** 12 more than −5
(g) 6 less than −1	**(h)** 20 more than −5
(i) 20 less than −30	**(j)** 100 more than −50
(k) 5 less than −100	**(l)** 200 less than 0
(m) 14 more than −12	**(n)** 5 bigger than −4
(o) 6 bigger than −7	**(p)** 10 smaller than −3
(q) 150 smaller than −50	**(r)** 200 bigger than −300.

> Use a number line to help you.

5 The temperature in Belfast at 6 am on Christmas Day was −3 °C. By noon on Christmas Day the temperature in Belfast had risen by 9 °C. Work out the temperature in Belfast at noon on Christmas Day.

6 The temperature at the top of a mountain is −8 °C. The temperature at the bottom of this mountain is 20 °C higher than the temperature at the top. Work out the temperature at the bottom of the mountain.

7 When an aeroplane takes off, the temperature at ground level is 15 °C. The aeroplane flies to a height of 9000 metres. At this height the temperature is 35 °C lower than the temperature at ground level. Work out the temperature at 9000 metres.

8 The highest point of a tree is 6 metres above ground level.
The lowest point of the roots of the tree is 8 metres below
ground level. Work out the distance from the lowest point
of the roots of the tree to the highest point of the tree.

9 The temperature at the top of a mountain is −14 °C.
The temperature at the bottom of the mountain is 12 °C.
Work out the difference between the temperature at the
bottom of the mountain and the temperature at the top of
the mountain.

10 On December 19th the difference
between the temperatures in
Moscow and Luxor was 31 °C.
Luxor was warmer than Moscow.
The temperature in Moscow on
December 19th was −10 °C.
Work out the temperature in Luxor
on December 19th.

Exercise 2.2   Links 2B–F

For questions **1** to **3** use the number line from −10 °C to
+10 °C to help you.

1 Find the number of degrees between each pair of
temperatures.

   (a) 2 °C, −3 °C   (b) 9°C, −1 °C   (c) −3 °C, 5 °C

   (d) 8 °C, −5 °C   (e) −4 °C, −10 °C   (f) 0 °C, −8 °C

   (g) 1 °C, 8 °C   (h) −1 °C, 9 °C

2 Find the temperature when

   (a) 0 °C rises by 4 °C   (b) −3 °C falls by 5 °C

   (c) 7 °C falls by 12 °C   (d) −2 °C rises by 4 °C

   (e) 5 °C falls by 9 °C   (f) −1 °C rises by 6 °C

3 Rearrange the temperatures in each part in order, lowest
temperature first.

   (a) −3 °C, 5 °C, 7 °C, −1 °C, 0 °C, 3 °C, −7 °C, −4 °C

   (b) 9 °C, −4 °C, −9 °C, 4 °C, −3 °C, 2 °C, −2 °C, 3 °C

   (c) −1 °C, −10 °C, −5 °C, 5 °C, 3 °C, 10 °C, 8 °C, −3 °C

°C
+10
+9
+8
+7
+6
+5
+4
+3
+2
+1
0
−1
−2
−3
−4
−5
−6
−7
−8
−9
−10

4 Work out these additions and subtractions.

(a) $-5--5$ (b) $+7+-3$ (c) $-9-+6$

(d) $+2++2$ (e) $-7+-3$ (f) $+14--8$

(g) $-1+-4$ (h) $+5--7$ (i) $9-0$

(j) $+9+-10$ (k) $-7++13$ (l) $+3--8$

(m) $-9--16$ (n) $-4-+14$ (o) $+15++8$

(p) $-9-+13$

> **Remember:** When adding, subtracting, multiplying or dividing: two like signs give a +
> $++=+$
> $--=+$
> two unlike signs give a $-$
> $-+=-$
> $+-=-$

5 Work out these mutiplications and divisions.

(a) $+2\times+3$ (b) -2×-4 (c) $-3\times+4$

(d) $+5\times-3$ (e) $+15\div-3$ (f) $+16\div+4$

(g) $-20\div-2$ (h) $+9\div-1$ (i) -6×-7

(j) -8×-4 (k) $+6\times-4$ (l) $+24\div-3$

(m) $-30\div-10$ (n) $+9\times-7$ (o) $-48\div+4$

(p) $-28\div-2$ (q) $+9\times-3$ (r) -7×-4

(s) $+96\div-6$ (t) $-56\div-8$

Exercise 2.3   Link 2G

1 (a) Write down all the even numbers from the list.

   4, 9, 20, 56, 73, 115, 906, 1002, 43 278

(b) Write down all the odd numbers from the list.

   4, 9, 23, 56, 74, 218, 827, 2001, 56 983

2 (a) Write down all the prime numbers between 20 and 30.

(b) Write down the smallest even number greater than 50.

(c) Write down the largest odd number less than 50.

(d) Write down the smallest prime number greater than 50.

3 Write down all the factors of

(a) 8   (b) 24   (c) 100   (d) 280

> The factors of 15 are
> 1, 3, 5, 15
> The multiples of 4 are
> 4, 8, 12, 16, 20, ...
> The first six prime numbers are
> 2, 3, 5, 7, 11, 13

4 Find all four factors of the number 111.

5 Write down the first five multiples of

(a) 3   (b) 10   (c) 12   (d) 200

6 Find the number which is a multiple of 5 and also a multiple of 8.

7 Find the first multiple of 9 which is greater than 50.

8 Find the number which is a common factor of

   (a) 12 and 15   **(b)** 14 and 63

   (c) 15 and 50   **(d)** 33 and 77.

9 Find the numbers which are common factors of

   (a) 20 and 50   **(b)** 42 and 105.

10 Copy these three rows of numbers.

       1 2 3 4 5 6 7 8 9 10
    11 12 13 14 15 16 17 18 19 20
    21 22 23 24 25 26 27 28 29 30

   (a) Draw a ○ around each multiple of 3.

   (b) Draw a × through each multiple of 4.

   (c) Write down all the numbers less than 30 which are both a multiple of 3 and a multiple of 4.

   (d) Find the first number greater than 30 which is a multiple of both 3 and 4.

Exercise 2.4   **Links 2H–J**

1 The first three square numbers are shown as dot patterns.

   (a) Draw the dot pattern for the 4th square number.

   (b) Work out the 4th square number.

   (c) Work out the 5th square number.

   1st = 1   2nd = 4   3rd = 9

2 Use your calculator to find

   (a) the 5th square number

   (b) the 3rd cube number

   (c) the 18th square number

   (d) the 9th cube number

   (e) the 8th to 15th square numbers

   (f) the 10th to 13th cube numbers.

> Square numbers are numbers that are multiplied by themselves
> e.g. $2 \times 2 = 4$.
> Cube numbers are numbers that are multiplied by themselves twice
> e.g. $2 \times 2 \times 2 = 8$
> $\sqrt{16} = 4$ $\sqrt[3]{8} = 2$

3 From this list write down all the numbers which are

   (a) square numbers

   (b) cube numbers.

   100, 27, 1, 343, 625, 324, 96, 1331, 32, 125, 256

4 Use your calculator to work out the following.
   Give your answers correct to 3 significant figures.

   (a) 14^2 (b) 4.6^2 (c) 36^3

   (d) 17.2^2 (e) 0.012^3 (f) 13.9^2

   (g) $(-4.3)^3$ (h) $\sqrt{361}$ (i) $\sqrt[3]{729}$

   (j) $\sqrt[3]{1728}$ (k) $\sqrt{1296}$ (l) $\sqrt{0.0196}$

> To round to 3 significant figures (3 s.f.), look at the place value of the 3rd digit from the left and round the number to this place value.

5 Use your calculator to work out the following.
   Give your answers correct to 2 decimal places.

   (a) $\sqrt{198}$ (b) $\sqrt{3479}$ (c) $\sqrt{21.9}$

   (d) $\sqrt{392\,761}$ (e) $\sqrt[3]{10}$ (f) $\sqrt[3]{-14.9}$

   (g) $\sqrt[3]{47.8}$ (h) $\sqrt[3]{1219.7}$ (i) $\sqrt[3]{7\,846\,591}$

6 Use a trial and improvement method to find these roots
   correct to 2 decimal places. Use a calculator to check your
   answers.

   (a) $\sqrt{6}$ (b) $\sqrt[3]{19}$ (c) $\sqrt{19}$

   (d) $\sqrt[3]{32}$ (e) $\sqrt{60}$ (f) $\sqrt[3]{48}$

   (g) $\sqrt{39}$ (h) $\sqrt[3]{40}$ (i) $\sqrt{27}$

Exercise 2.5   Links 2K, L

1 Write these numbers as products of their prime factors.

   (a) 50   (b) 300   (c) 34   (d) 78

> **Remember:** Prime factor form is writing a number as the product of its prime factors, e.g. 24 in prime factor form is $2 \times 2 \times 2 \times 3 = 2^3 \times 3$

2 Write these numbers in prime factor form.

   (a) 12   (b) 36   (c) 54   (d) 17

3 Find the highest common factor of

   (a) 4 and 12   (b) 16 and 24   (c) 14 and 70

   (d) 42 and 28   (e) 72 and 30   (f) 8 and 12

   (g) 21 and 70   (h) 2 and 8.

> To find the highest common factor (HCF) of 48 and 60:
> Factors of 48:
> $2 \times 2 \times 2 \times 2 \times \underline{3}$
> Factors of 60:
> $2 \times 2 \times \underline{3} \times 5$
> $2 \times 2 \times 3 = 12$
> So HCF is 12.

4 Find the lowest common multiple of

   (a) 2 and 3   (b) 3 and 5   (c) 8 and 12

   (d) 12 and 18   (e) 48 and 60   (f) 10 and 25

   (g) 18 and 60   (h) 24 and 60.

> To find the lowest common multiple, look at the multiples of each number in turn and find one that is in both lists.

3 Essential algebra

Exercise 3.1
Links 3A–D

1 Use algebra to write
  (a) 4 more than a (b) 3 less than b
  (c) c with 5 added   (d) d more than c
  (e) $4e$ with 7 subtracted.

2 Write these in short form.
  (a) $x + x + x$ (b) $y + y$
  (c) $b + b + b + b + b$ (d) $w + w + w + w$
  (e) $p + p + p + p + p + p + p + p$ (f) $q + q + q + q + q + q$

3 Write these out in a longer form.
  (a) $3y$ (b) $4x$ (c) $5w$ (d) $2z$
  (e) $5a$ (f) $7c$ (g) $9d$ (h) $4y$

4 Make these expressions simpler by adding or subtracting like terms.
  (a) $2x + 3x$ (b) $3y + y$ (c) $5a - 2a$
  (d) $4w + 3w$ (e) $5a + 4a + 3a$ (f) $8b - 3b$
  (g) $5x + 6x - 3x$ (h) $2d + 4d$ (i) $5y - 3y + 2y$
  (j) $12x - x$ (k) $10a - 10a$ (l) $9s - 9s + s$

Remember
a means $1a$
$2a + a = 3a$

5 Simplify these expressions completely by collecting like terms.

Remember: The $+$ or $-$ sign is part of each term.

  (a) $3y + 2x + 4y + 5x$ (b) $6a + 2b + 5a + 5b$
  (c) $3d + 2f + 3d$ (d) $3s + 2t - s + 2t$
  (e) $4d + d + 3d - 5d$ (f) $5x - 2y + 3x + 2x$
  (g) $4p + 5p + 6p - 9p - 5p$ (h) $6a + 2a - 3a + 4a - 4a$
  (i) $3x + 1 + 3x + 1 + 2$ (j) $5 + 2a + 3a - 2a + 4$
  (k) $4a + 7b - 2a - 6b - b$
  (l) $4y - 3x + 2 + 3x - 4y + 2$
  (m) $5b - 3b + 1 - 7b - 1 - 2b$
  (n) $3p + 7p - 10p + 7 + p$
  (o) $4x + 3y + 2y + 5x - 3x + 2y$
  (p) $8a + 4a - 5a + 2a - 7a + 3a$

Exercise 3.2

Use multiplication signs to write the expressions out in a longer form.

1	ab	**2**	cde	**3**	$2a$	**4**	$3ab$
5	$5xyz$	**6**	$10abc$	**7**	$7mnp$	**8**	$4abcd$
9	$15xyz$	**10**	$9abcd$	**11**	$3xyz$	**12**	$21abc$

> **Remember:**
> ab means $a \times b$.

> **Remember:**
> $11bc$ is $11 \times b \times c$ not $1 \times 1 \times b \times c$.

Write these expressions in a simpler form.

13	$a \times b$	**14**	$x \times y \times z$	**15**	$3 \times w \times s \times t$
16	$5 \times m \times m$	**17**	$4 \times b \times c \times d$	**18**	$h \times k \times l$
19	$2 \times s \times t$	**20**	$7 \times a \times b \times d$	**21**	$s \times 5 \times t \times w$
22	$2x \times 3y$	**23**	$4a \times 5b$	**24**	$5a \times 2b$
25	$a \times b \times c$	**26**	$6s \times 2t$	**27**	$8y \times 3z$
28	$2a \times 2b \times 2c$	**29**	$5g \times 2f$	**30**	$9a \times 3b \times c$
31	$10x \times 9y \times 3z$	**32**	$5d \times 3e$	**33**	$e \times 2f \times g$
34	$6a \times 4b \times 2c$	**35**	$4a \times 4b \times 4c$	**36**	$11a \times 10b \times 3c$
37	$16x \div 4$	**38**	$15y \div y$	**39**	$14ab \div 2a$
40	$9xyz \div 3xz$	**41**	$\dfrac{64pq}{8}$	**42**	$\dfrac{32a}{8a}$
43	$\dfrac{24b}{16}$	**44**	$\dfrac{2xy}{16x}$	**45**	$\dfrac{ab \times 12c}{3b}$

> **Remember:** Numbers come first then letters in alphabetical order.

Exercise 3.3

Work out the expressions in questions **1** to **3**.

1 (a) 3^3 (b) 2^5 (c) 5^6

2 (a) 11^3 (b) 7^3 (c) 10^5

3 (a) 7^5 (b) $5^3 + 4^2$ (c) $12^2 \div 3^4$

Find the value of x in questions **4** to **6**.

4 (a) $2^x = 16$ (b) $4^x = 64$ (c) $5^x = 625$

5 (a) $2^x = 128$ (b) $7^x = 343$ (c) $4^x = 1024$

6 (a) $10^x = 10\,000\,000$ (b) $8^x = 4096$ (c) $6^x = 7776$

Simplify by writing as a single power of the number.

7 (a) $2^5 \times 2^3$ (b) $6^4 \times 6^6$ (c) $4^3 \times 4^7$

8 (a) $2^7 \div 2^4$ (b) $5^6 \div 5^1$ (c) $3^7 \div 3^6$

> **Remember these rules:**
> $a \times a \times a \times a = a^4$
> a can be written as a^1
> $x^a \times x^b = x^{a+b}$
> $x^a \div x^b = x^{a-b}$
> $(x^a)^b = x^{ab}$

9 (a) $2^3 \times 2^4 \times 2^6$ (b) $7^2 \times 7^5 \times 7^4$ (c) $7^5 \div 7^1$

10 (a) $\dfrac{5^5 \times 5^4}{5^6}$ (b) $\dfrac{2^8 \times 2^5}{2^7}$ (c) $\dfrac{7^4 \times 7^6}{7^9}$

Exercise 3.4   Links 3J–L

1 Write these expressions using powers.
   (a) $x \times x \times x$
   (b) $y \times y$
   (c) $a \times a \times a \times a \times a$
   (d) $s \times s \times s$
   (e) $b \times b \times b \times b \times b$
   (f) $p \times p \times p$

2 Write these expressions in full.
   (a) p^2
   (b) q^5
   (c) r^3

3 Simplify
   (a) $x^3 \times x^2$
   (b) $y^4 \times y^5$
   (c) $a^3 \times a^5$
   (d) $a \times a^2$
   (e) $(b^2)^3$
   (f) $x^6 \div x^2$
   (g) $b^8 \div b^3$
   (h) $c^{20} \div c^{14}$
   (i) $n^2 \div n$
   (j) $(b^4)^5$
   (k) $(a^3)^0$
   (l) $x^2 \times x^7$
   (m) $4x^3 \times 3x^2$
   (n) $5p \times 2p^3$
   (o) $2y^{10} \times y^3$
   (p) $6x^3 \div 2x$
   (q) $24p^7 \div 8p^3$
   (r) $12a^6 \div 3a^2$
   (s) $(3a^2)^3$
   (t) $(5b^7)^2$
   (u) $\dfrac{4x^3 \times 6x^2}{3x^4}$

Exercise 3.5   Links 3M–S

1 Use BIDMAS to help find the value of these expressions.
   (a) $4 + (5 + 2)$
   (b) $6 - (4 + 1)$
   (c) $6 \times 3 + 2$
   (d) $25 \div 5 + 5$
   (e) $11 - (4 - 3)$
   (f) $30 \div 5 - 3$
   (g) $(4 + 5)^2$
   (h) $2 \times (3 + 4)^2$
   (i) $3^2 + 4^2$
   (j) $\dfrac{6^2 - 4^2}{5}$
   (k) $\dfrac{(1 + 3)^2}{2^2 - 2}$
   (l) $2 \times 3^2 + 3 \times 4^2$

> **Remember:**
> BIDMAS order
> Brackets
> Indices
> × and ÷
> + and −

2 Expand the brackets in these expressions.
   (a) $4(a + b)$
   (b) $5(x - y)$
   (c) $2(3x + 2y)$
   (d) $3(2x - 3y)$
   (e) $7(2x - y + 3z)$
   (f) $11(3a + 9b)$
   (g) $4(x + y) + 3(x + y)$
   (h) $5(a - b) + 6(2a + 3b)$
   (i) $2(p + 3q) + 5(2p - q)$
   (j) $7(s + 2t) + 3(2s - 3t)$

3 Work out
   (a) $15 - (4 + 2)$
   (b) $18 - (5 + 3)$
   (c) $8 - (6 - 5)$
   (d) $13 - (7 + 6)$
   (e) $9 - (8 - 3)$
   (f) $21 - (8 - 5)$

4 Write these expressions as simply as possible.

Expand the brackets and collect like terms.

   (a) $3x + 4y - (x + y)$ **(b)** $2a + 5b - (a + b)$

   (c) $5s + 4t - (3s - 2t)$ **(d)** $3(4w + 5x) - 2(3w + 2x)$

   (e) $4(x + 2y) - 3(x + 4y)$

   (f) $5(2a + 3b - 4c) - 3(3a + 2b - 2c)$

   (g) $3(2p - 3q) + 2(3p + 2q) - 4(2p - q)$

5 Expand the brackets in the following expressions.

   (a) $3(x + 2)$ **(b)** $5(2p + 1)$ **(c)** $4(x - 3)$

   (d) $2(3p - 2)$ **(e)** $3(a + 2b)$ **(f)** $9(2n - 5)$

   (g) $x(2x + 5)$ **(h)** $a(x + a)$ **(i)** $p(3p + 1)$

   (j) $3x(2x - 1)$ **(k)** $n(n + 2)$ **(l)** $4b(b + 2)$

   (m) $2t^2(t + 3)$ **(n)** $5x(2 - x)$ **(o)** $4y^3(2y - 3y^2)$

Exercise 3.6   Link 3T

1 Factorise each of these expressions.

   (a) $3x + 6$ **(b)** $4x - 12$ **(c)** $25p + 5$

   (d) $8a + 4$ **(e)** $ab + a$ **(f)** $3x + 12y$

   (g) $4a + 12b$ **(h)** $120x + 6$ **(i)** $3x + 6y + 12z$

   (j) $3xy + 9x$ **(k)** $10 - 5y$ **(l)** $7a + 14b$

   (m) $3 - 9x$ **(n)** $12p - 3q$ **(o)** $250n + 25m$

2 Factorise each of these expressions.

Look for a common factor - there may be more than one.

   (a) $5x^2 + 4x$ **(b)** $3y - 6y^2$ **(c)** $4b^2 + b$

   (d) $3z - z^2$ **(e)** $p^2 - 2p$ **(f)** $n - n^2$

   (g) $m^2 + 2m$ **(h)** $x^2 - 3x$ **(i)** $x^3 - 2x$

   (j) $a^3 + a$ **(k)** $2x^3 + 3x$ **(l)** $x^2 - 4x$

   (m) $4y^4 + 3y$ **(n)** $m^2 - m$ **(o)** $2t + 3t^2$

3 Factorise completely.

   (a) $6x^2 + 3x$ **(b)** $4y - 8y^2$ **(c)** $3x - 15x^2$

   (d) $7p^2 - 2p$ **(e)** $7b^2 + 3b$ **(f)** $5y - 3y^2$

   (g) $4x + 3x^2$ **(h)** $4xy + 12y$ **(i)** $x^3 + 6x^2$

   (j) $ab + a^2$ **(k)** $24x^3 - 18x$ **(l)** $ax - a$

   (m) $xy + 3x$ **(n)** $x^2y + xy$ **(o)** $a^2x + 3ax$

   (p) $6p^2 - 9p$ **(q)** $a^2b + ab$ **(r)** $p^2q + 2pq^2$

   (s) $4x^3y + 6xy$ **(t)** $3pq^2 + 2pq$ **(u)** $\pi r^2 + 2\pi rh$

   (v) $2x^3y^2 + 4xy$ **(w)** $ut + gt^2$ **(x)** $6a^2b^3 + 15ab^4$

4 Patterns and sequences

Exercise 4.1

Links 4A, B

Find the two missing numbers in these number patterns.
Write down the rule for each pattern too.

1   4, 6, 8, —, —, 14, 16
2   5, 8, 11, 14, —, —, 23, 26
3   9, 19, —, —, 49, 59, 69
4   17, 21, —, —, 33, 37, 41
5   29, 25, 21, —, —, 9, 5
6   98, 88, 78, —, —, 48, 38
7   17, 15, —, —, 9, 7, 5
8   80, 71, 62, —, —, 35, 26

> You need to find a rule that adds or subtracts the same amount each time for each question.

Exercise 4.2

Links 4C, D

Find the two missing numbers in these number patterns.
Write down the rule for each pattern too.

1   3, 6, 12, —, —, 96, 192
2   2, 6, 18, —, —, 486, 1458
3   1, 6, —, —, 1296, 7776
4   11, 110, —, —, 110 000, 1 100 000
5   64, 32, —, —, 4, 2
6   729, 243, —, —, 9, 3, 1
7   3072, 768, —, —, 12, 3
8   12 500, 2500, —, —, 20, 4

> You need to multiply or divide by the same number each time for each question.

Exercise 4.3

Link 4E

Copy the pattern, find the difference and the rule for each one, and the next number.

1   5, 7, 9, 11, 13, ...
2   3, 8, 13, 18, 23, ...
3   2, 4, 6, 10, 16, ...
4   1, 3, 4, 7, 11, ...
5   1, 4, 9, 16, 25, ...
6   3, 3, 6, 9, 15, ...
7   1, 6, 11, 16, 21, ...

> Questions 3, 4 and 6 are Fibonacci type patterns.

Exercise 4.4

Link 4F

1

 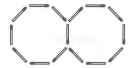 

8 matches   15 matches   22 matches

(a) Draw the next two patterns.

(b) Copy and complete the table.

(c) Write down the rule to find the number of matches in the next pattern.

(d) Find the general rule for the nth term.

Term number	1	2	3	4	5
Matches used	8	15	22		

2 For these patterns

(a) draw the next two patterns

(b) write down the rule then find the next pattern

(c) find the nth term in the pattern

(d) use your rule to find the 10th term.

(i) ×   × ×   × × ×   × × × ×

(ii)
```
                    ×
          ×         ×
×        × ×       × × ×
```

(iii)

> For a pattern that is getting smaller and smaller the term in n is negative, e.g. $43 - 2n$.

3 Find the general rule for the nth term for each pattern. Then use your rule to find the 20th term.

(a) 2, 4, 6, 8, 10, 12, ...

(b) 4, 8, 12, 16, 20, 24, ...

(c) 9, 18, 27, 36, 45, 54, ...

(d) 3, 6, 9, 12, 15, 18, ...

(e) 4, 6, 8, 10, 12, 14, ...

(f) 39, 35, 31, 27, 23, 19, ...

(g) 0, 7, 14, 21, 28, 35, ...

(h) 39, 37, 35, 33, 31, 29, ...

(i) 1, 4, 7, 10, 13, 15, 18, ...

(j) 100, 90, 80, 70, 60, 50, ...

Exercise 4.5

Link 4G

For each number pattern, explain whether the numbers in brackets are members of the number pattern.

1 1, 4, 7, 10, 13, 15, 18, ...   (28, 33)

2 2, 6, 10, 14, 18, 22, ...   (38, 97)

3 50, 44, 38, 32, 26, ...   (18, 16)

4 3, 9, 15, 21, 27, ...   (85, 93)

5 81, 77, 73, 69, 65, ...   (41, 35)

5 Decimals

Exercise 5.1

Links 5A, B

1 Draw a copy of the place value diagram. On your copy of the place value diagram write in these numbers:

(a) 34.7   (b) 3.47   (c) 0.347

(d) 243.75   (e) 0.072   (f) 132.89

(g) 10.003   (h) 0.04   (i) 20.02

(j) 0.001   (k) 3.01   (l) 25.67

2 What is the place value of the underlined digit in each number?

(a) 4$\underline{7}$.9   (b) 4.$\underline{7}$9   (c) 0.4$\underline{7}$9   (d) 2$\underline{5}$3.81

(e) 18.1$\underline{0}$6   (f) 24.03$\underline{8}$ (g) 110.3$\underline{2}$ (h) 0.07$\underline{1}$

(i) $\underline{2}$3.4   (j) 0.32$\underline{1}$ (k) 2$\underline{0}$.03   (l) $\underline{2}$45.672

3 The table gives the heights in metres of six people.

Steven	Gemma	Fatima	Paul	Yvonne	Yasmin
1.83	1.57	1.63	1.85	1.68	1.75

Write the list of names in order of height, starting with the shortest.

4 Re-arrange these decimals in order of size, starting with the smallest.

(a) 4.73, 7.43, 3.47, 4.07, 7.04

(b) 15.03, 1.503, 5.103, 3.105, 1.035, 31.05

(c) 0.0011, 10.10, 0.1001, 0.1100

(d) 1.061, 1.106, 1.601, 1.016

(e) 0.07, 0.70, 0.007, 7.00

(f) 5.16, 5.016, 5.061, 5.61

> Look for the smallest number in each row.

5 Re-arrange these decimals in order of size. Start with the largest.

(a) 4.4, 4.12, 4.75, 7.45, 5.74

(b) 0.011, 0.0011, 0.101, 0.110

(c) 0.52, 0.075, 0.507, 0.4, 0.08

(d) 0.07, 0.14, 0.008, 0.205, 0.025

(e) 3.09, 2.08, 3.2, 2.3, 2.267

(f) 2.222, 22.22, 222.2, 2222, 2.222 22

6 Eight people took part in a 400 metres race. Their names and times, in seconds, for the race are given below.

Smith	104.82	Thomas	102.37
Warne	103.41	Kelly	103.47
Khan	102.15	McGowan	104.72
Young	105.28	Priest	103.94

The winner is the person with the lowest time.

(a) Who won the race?

(b) Who came last?

> **Remember:**
> *Decimal places*
> Count the number of digits after the decimal point.
>
> *Significant figures*
> Start counting when you reach the first digit that is not zero.

Exercise 5.2   Links 5C–E

1 Round
(a) 18.6 correct to the nearest whole number
(b) 102.48 correct to the nearest whole number
(c) 3.53 correct to one place of decimals
(d) 14.365 correct to one place of decimals
(e) 25.057 correct to two places of decimals
(f) 0.053 correct to one place of decimals.

2 Write each number to **(i)** 1 s.f. **(ii)** 2 s.f. **(iii)** 3 s.f.
(a) 1950   (b) 2.550   (c) 5608
(d) 250 250   (e) 9.999   (f) 3 250 000
(g) 0.002 562   (h) 0.000 750 9   (i) 12.005
(j) 100.05   (k) 26.895   (l) 50.999

3 The winning time for a 200 metres race was 23.468 seconds. Round this number correct to 3 significant figures.

4 The engine capacity, in litres, of Isabella's car is 1.398. Round this number correct to
(a) the nearest whole number of litres
(b) one place of decimals
(c) three significant figures.

Exercise 5.3

1 Work these out **without a calculator**, showing all your working.
   (a) 2.7 + 5.6   (b) 0.75 + 4   (c) 32.6 + 53.2
   (d) 324.8 + 2.25   (e) 349 + 34.9   (f) 0.017 + 1.235
   (g) 62 + 6.2   (h) 143.5 + 3.5   (i) 5.67 + 0.033

2 Work these out **without a calculator**, showing all your working.
   (a) 7.4 + 17.6 + 27.3   (b) 23.07 + 8.6 + 7.44
   (c) 0.723 + 1.1 + 12.453   (d) 3.8 + 3.8 + 3.8
   (e) 4.07 + 32 + 0.0457   (f) 214.073 + 92 + 5.91

3 Work these out **without a calculator**, showing all your working.
   (a) 28.8 − 17.3   (b) 6.72 − 2.6   (c) 32.8 − 8.2
   (d) 100 − 8.5   (e) 0.68 − 0.52   (f) 15 − 2.85
   (g) 1000 − 986.4   (h) 0.17 − 0.073 (i) 305.07 − 56.28

4 Simon earns £40 a week at his part-time job.
   He pays his mother £4.50 for driving him to and from work.
   How much does he have left?

5 Sandra earns £125 a week. She pays tax of £10.50 and
   National Insurance of £7.80.
   How much money does she have left to take home?

6 Cherie takes part in a three-hour endurance race. She runs
   8.4 km in the first hour, 7.8 km in the second hour and
   9.4 km in the third hour. How far does she run in total?

7 Bobbi goes shopping with a £20 note. She buys a magazine for
   £2.75, a top for £7.99 and some shampoo for £3.99.
   How much of the £20 does she have left?

8 Tom buys some new trainers for £85.99. He pays with five
   £20 notes. How much change should he receive?

Exercise 5.4

1 Work these out **without a calculator**, showing all your working.
   (a) The total cost of 5 CDs at £2.99 each.
   (b) The total cost of 6 cans of drink at £0.45 each.
   (c) The total cost of 12 litres of petrol at £0.74 per litre.
   (d) The total cost of 2.5 kilograms of apples at £1.18 per
       kilogram.

2 Work these out **without a calculator**, showing all your working.

(a) 8.7×4 (b) 0.87×4 (c) 0.87×40

(d) 1.3×10 (e) 23.5×12 (f) 32.5×22

(g) 71.3×10 (h) 2.1×5 (i) 0.6×0.8

(j) 0.35×0.6 (k) 1.7×0.3 (l) 0.1×0.1

(m) 0.04×0.2 (n) 0.03×0.08 (o) 3.14×0.6

(p) 0.4×0.4

3 Work these out **without a calculator**, showing all your working.

(a) $81.6 \div 3$ (b) $27 \div 10$ (c) $47.5 \div 5$

(d) $24.68 \div 4$ (e) $7.8 \div 20$ (f) $54.3 \div 6$

(g) $37.8 \div 10$ (h) $0.46 \div 10$ (i) $57 \div 100$

(j) $3 \div 100$ (k) $0.38 \div 100$ (l) $75 \div 1000$

4 Eight people share £194.80 equally.
How much will each person receive?

5 How many 4 gallon tanks will be needed to hold 78.4 gallons of oil?

6 Work these out **without a calculator**, showing all your working.

(a) $4 \div 0.2$ (b) $8.64 \div 2.4$ (c) $19.2 \div 12.8$

(d) $0.258 \div 0.3$ (e) $1.7 \div 0.5$ (f) $58.08 \div 12.1$

(g) $0.003 \div 0.15$ (h) $3.2 \div 0.8$ (i) $360.4 \div 6.8$

(j) $0.035 \div 0.01$ (k) $10.1 \div 0.4$ (l) $2.624 \div 6.4$

7 Louis buys 8 DVDs at £12.99 each.
How much does this cost him?

8 Robbie has 8 litres of cola.
How many glasses, each holding 0.25 litres, can he fill?

9 Gareth buys 82.5 litres of diesel for his boat.
The diesel costs 97p per litre.
How much does the diesel cost him?
Give your answer in pounds and pence.

10 Susan moves 5 tonnes of manure. She uses a wheelbarrow that holds 25 kg.
How many 25 kg loads does she need to move?

6 Angles and turning

Exercise 6.1

Link 6A

1 Write down which of these are turning movements.
   (a) Using a door handle to open a door.
   (b) Pulling a bucket up a well.
   (c) Winding up a clockwork toy.
   (d) A stone dropped from a cliff.
   (e) The minute hand of a clock as time passes.
   (f) Using a remote control to turn the TV on.

2 Here is a map of a village.

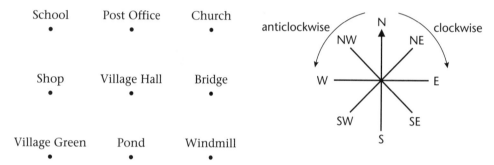

Peter is standing at the village hall. He is facing North.
   (a) After a quarter turn clockwise what is he facing?
       Which compass direction is he facing?

Again Peter is standing at the village hall. He is facing North.
   (b) After a half turn anticlockwise what is he facing?
       Which compass direction is he facing?

Mary is at the pond facing towards the shop.
   (c) What direction is she facing?
   (d) After a three-quarters turn anticlockwise what is she facing?
   (e) What direction is she facing now?

3 Look at the map in question 2. What direction is
   (a) the church from the school
   (b) the village green from the school
   (c) the shop from the post office
   (d) the post office from the bridge
   (e) the windmill from the school?

4 How much does the hour hand of a clock turn between

(a) 2 pm and 8 pm (b) 4 pm and 10 pm

(c) 10 am and 1 pm (d) 1 am and 10 am

(e) 3 am and 3 pm (f) 9:30 am and 9:30 pm?

Exercise 6.2

Links 6B, C

In questions **1** to **6**, name each marked angle and say whether it is acute, obtuse or right-angled.

1

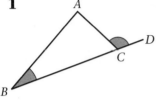

2

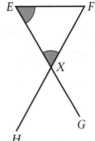

3

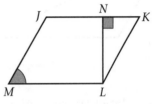

4

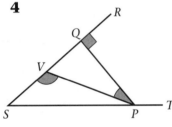

5

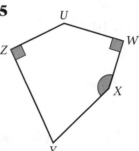

> **Remember:**
> Acute angles are less than 90°
> Obtuse angles are between 90° and 180°
> Reflex angles are between 180° and 360°
> Right angles are exactly 90°

6 From the diagrams above, estimate the size of these angles.

(a) *BAC* (b) *HXF* (c) *NKL*

(d) *NLM* (e) *VSP* (f) *QPT*

7 Draw a quadrilateral with two obtuse angles.

Exercise 6.3

Links 6D, E

1 Measure the angles as accurately as you can.

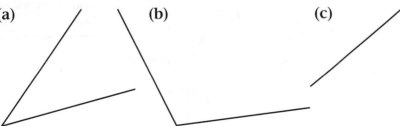

(a) (b) (c)

> Use a straight edge like a piece of paper to extend the lines if you need to.

2 Measure all three angles in each of these triangles.

(a) **(b)**

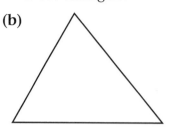

(c) **(d)**

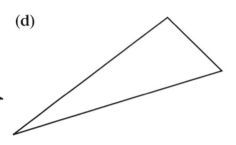

3 Draw and label these angles.

(a) *ABC* = 20°   (b) *DEF* = 40°

(c) *GHI* = 60°   (d) *JKL* = 35°

(e) *MNO* = 75°   (f) *PQR* = 105°

(g) *STU* = 77°   (h) *VWX* = 13°

(i) *YZA* = 129°   (j) *BCD* = 146°

Exercise 6.4   Links 6F, G

1 Calculate the marked angles in the diagrams below.
Give reasons for your answers.

(a)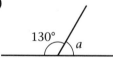

(b)

(c)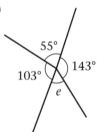

> **Remember:**
> The angles on a straight line add up to 180°
>
> The angles at a point add up to 360°
>
> The angles in a triangle add up to 180°
>
> The angles in a quadrilateral add up to 360°

(d)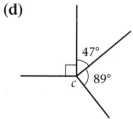

(e) **(f)**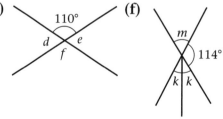

2 Make accurate drawings of these shapes.

(a)
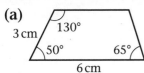
130°
3 cm
50° 65°
6 cm

(b)
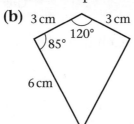
3 cm   3 cm
85° 120°
6 cm

(c)
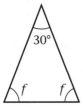
3 cm
50°
2.5 cm 108°
3 cm
3.5 cm

Lines marked with arrows are parallel

Exercise 6.5

Links 6H, I

1 Work out the lettered angles in these shapes.
Write down your reasons.

(a)

30°
f   f

(b)
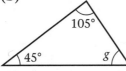
105°
45° g

(c)
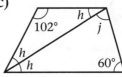
102° h
h   j
h   60°

(d)
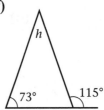
b
37°

(e)
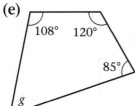
108° 120°
85°
g

(f)
h
73° 115°

2 Draw a copy of each diagram and find all the unmarked angles.
Write down your reasons.

(a)
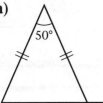
50°

(b)
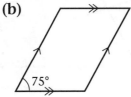
75°

(c)
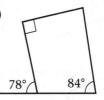
78° 84°

Remember: Sides marked in the same way are the same length.

(d)
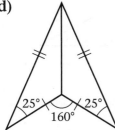
25° 25°
160°

(e)
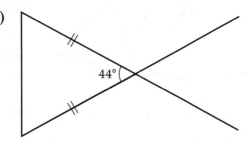
44°

Exercise 6.6 Link 6J

1 State the lines which are parallel to each other.

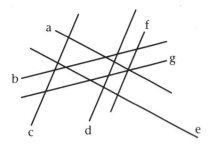

2 Copy this diagram and mark the parallel lines.

3 Find the size of the marked angles. Give reasons for your answers.

(a)

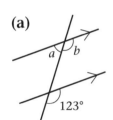

(b)

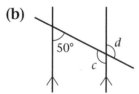

(c)

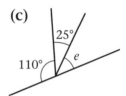

(d)

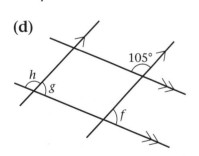

(e)

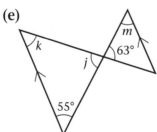

Remember:

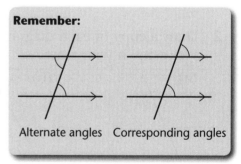

Alternate angles   Corresponding angles

(f)

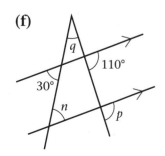

(g)

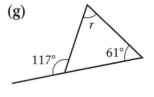

(h)

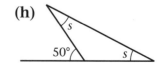

4 Calculate the named angles, giving reasons.

(a)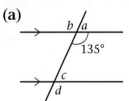

(b)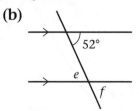

(c)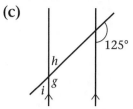

(d)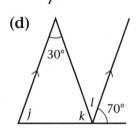

(e)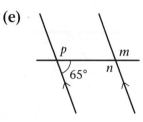

(f)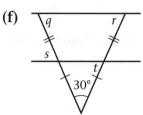

5 Calculate the named angles, giving reasons.

(a)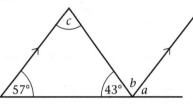

(b)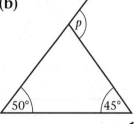

(c)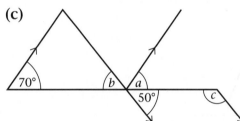

(d)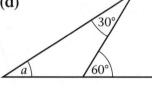

Exercise 6.7

Link 6K

1 Prove that angle *x* is 75°.

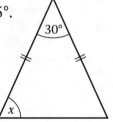

2 Prove that angle *y* is 115°.

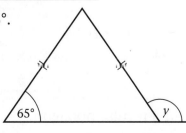

To prove something you have to explain <u>why</u> it is true.

3 Prove that triangle *ABC* is equilateral.

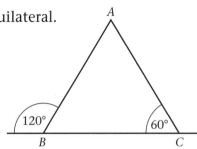

4 Prove that $x = 180° - a - b$.

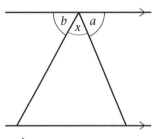

5 Prove that $a = c - b$.

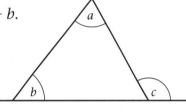

6 Prove that $x = \frac{1}{2}(180° - a)$.

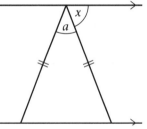

Exercise 6.8   Link 6L

1 Work out the exterior angle of a regular hexagon. Write down the interior angle of a regular hexagon.

2 A 20-sided shape is called an icosagon. Work out the exterior angle of a regular icosagon. Write down the sum of the interior angles of an icosagon.

3 A regular shape has an exterior angle of 20°. Work out how many sides it has. What is the size of an interior angle?

4 A nonagon has nine sides. Work out the interior angle for a regular nonagon.

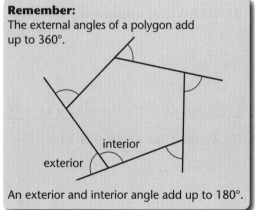

Remember:
The external angles of a polygon add up to 360°.

interior

exterior

An exterior and interior angle add up to 180°.

5 Find the number of sides of each regular polygon given the following:

(a) interior angle 144°
(b) exterior angle 24°
(c) exterior angle 40°
(d) exterior angle 30°
(e) sum of interior angles 3240°
(f) sum of interior angles 2880°

6 Find the sum of the interior angles of regular polygons with

(a) 16 sides (b) 25 sides (c) 32 sides (d) 36 sides.

7 Calculate the unknown angles in these polygons.

(a)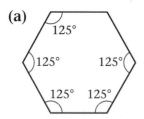
(b)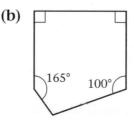
(c)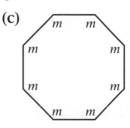

Exercise 6.9   Link 6M

1 Use a protractor to find the bearing of

(a) *B* from *A*
(b) *C* from *B*
(c) *A* from *C*.

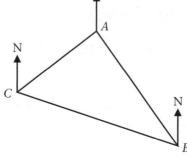

Remember: A bearing is an angle measured clockwise from facing North.

This is the bearing of *B* from *A*.

2 Work out the bearing of

(a) *Q* from *R*
(b) *P* from *R*
(c) *R* from *P*.

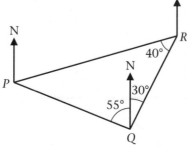

3 The bearing of Sheffield from Gloucester is 015°.
What is the bearing of Gloucester from Sheffield?

4 The bearing of York from Leeds is 060°.
What is the bearing of Leeds from York?

5 The bearing of Birmingham from Liverpool is 147°.
What is the bearing of Liverpool from Birmingham?

6 The bearing of Nottingham from Lincoln is 235°.
What is the bearing of Lincoln from Nottingham?

7 The bearing of Oxford from London is 290°.
What is the bearing of London from Oxford?

8 Write down the three-figure bearing for each of these directions.

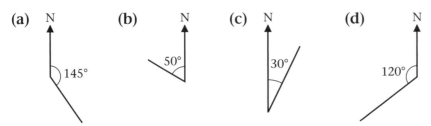

(a) N 145°   **(b)** N 50°   **(c)** N 30°   **(d)** N 120°

9 Write down the three-figure bearing for each of these directions.

(a) South   **(b)** West

(c) East   **(d)** South-West

(e) North-East   **(f)** North-West

(g) North   **(h)** South-East

10 Measure and write down the bearing of *B* from *A* in each of these diagrams.

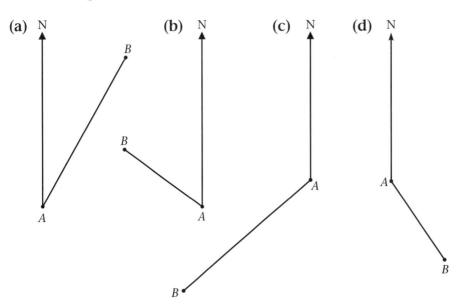

11 For each of the bearings in question **10**, work out the bearing of *A* from *B*.

7 2-D shapes

1 Complete the sentences using numbers, and words chosen from

       equal   parallel   opposite   sides   angles

   (a) A triangle has 3_____ and 3_____.

   (b) A quadrilateral has _____ sides.

   (c) _____ sides in a parallelogram are _____ and _____.

   (d) Pairs of adjacent _____ in a kite are _____.

   (e) The _____ of a square are equal.

   (f) An isosceles triangle has 2 _____ _____ and 2 _____ _____.

   (g) The _____ of a pentagon always add up to _____ degrees.

   (h) All the _____ of a rhombus are _____. The opposite _____ are _____ and _____.

2 Answer the questions about the 2-D shapes in the diagram.

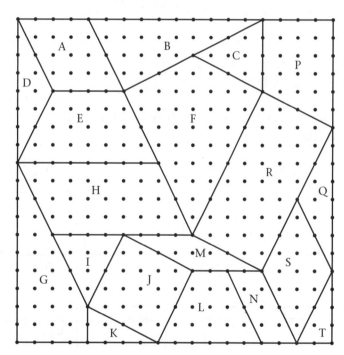

   (a) List the shapes which are parallelograms.

   (b) What mathematical name describes the shape R?

   (c) What mathematical name describes the shape J?

   (d) Which shape is a rhombus?

   (e) List the shapes which are trapeziums.

   (f) Which shapes are right-angled triangles?

   (g) Which shape is the kite?

3 On squared paper draw

    (a) three different triangles which have a base of 5 cm and a height of 3 cm

    (b) a trapezium with parallel sides 6 cm and 4 cm and with two angles as right angles

    (c) three different isosceles triangles with height 2 cm

    (d) a parallelogram with shortest sides 3 cm.

4 Investigate the number of different quadrilaterals you can make using 2, 3 or 4 identical right-angled triangles. In each case, give their mathematical name, e.g.

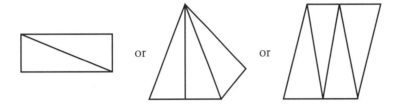

Exercise 7.2   **Link 7C**

In questions **1** to **4**, write down the letters of the shapes that are congruent.

> 'Congruent' means same shape and same size.

1

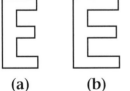 

  (a)   **(b)**   **(c)**   **(d)**   **(e)**

2

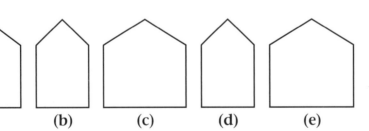

  (a)   **(b)**   **(c)**   **(d)**   **(e)**

3

  (a)   **(b)**   **(c)**   **(d)**   **(e)**

4

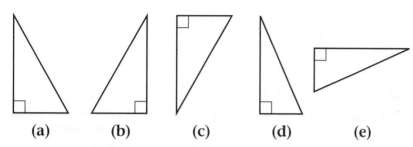

     (a)   (b)   (c)   (d)   (e)

5 List the pairs of congruent shapes in the diagram.

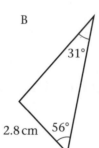

6 In each part, pick the two triangles which are congruent.

(a)

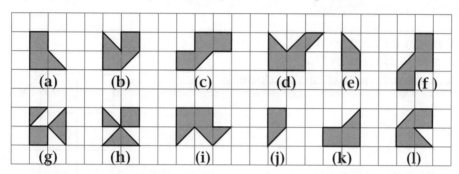

(b)

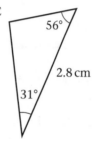

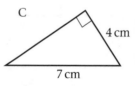

(c)

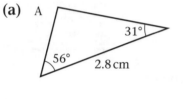

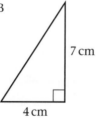

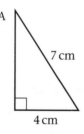

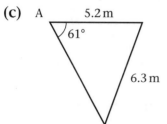

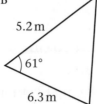

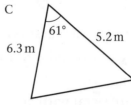

7 Draw a regular octagon.
Label the vertices *A*, *B*, *C*, *D*, *E*, *F*, *G*, *H*.
Join *AC*, *AD*, *CH* and *DH*.
Call the point where *AD* and *CH* cross *X*.
Name the pairs of triangles that are congruent.

Now join *DG*, *EG* and *EH*. Call the point where *DG* and *EH* cross *Y*. Make a list of the lines that are parallel. Identify and name quadrilaterals that are congruent.

> There should be four shapes with different mathematical names in your list.

Exercise 7.3   **Link 7D**

1 The shape shown can be used to tessellate the inside of a rectangle which is 6 squares by 9.
Show on squared paper how this can be done.   [E]

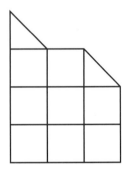

2 Show how these quadrilaterals tessellate.

(a)   **(b)**

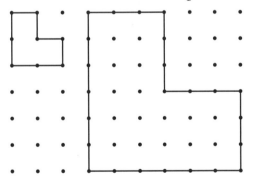

> **Remember:** Tessellate means covering all the space without gaps or overlapping.

3 Tessellate the small 'L' shape into the enlarged 'L' shape.

> The tessellation is not regular.

Exercise 7.4   **Link 7E**

1 Using ruler and compasses construct,
  (a) a triangle with sides 8 cm, 5 cm and 6 cm
  (b) a triangle with sides 7 cm, 6 cm and 6 cm
  (c) a quadrilateral with sides 6 cm, 8 cm, 13 cm and 13 cm where one of the diagonals is 10 cm long.

Using ruler, compasses and protractor make accurate drawings of the shapes in questions **2** to **6**.

First make a sketch so that you know what it looks like.

2 (a)

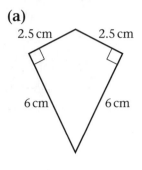

(b)

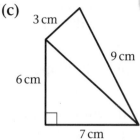

(c)

3 A triangle where $AB = 5$ cm, angle $BAC = 50°$ and angle $ABC = 65°$.

4 A right-angled triangle PQR with the right angle at R.
The lengths are $QR = 6$ cm and $PQ = 8.5$ cm.

5 A trapezium $WXYZ$ with WX parallel to YZ.
The lengths are $WX = 5$ cm, $ZY = 7$ cm and $WZ = 3$ cm.
Angle $WZY = 55°$.

6 A convex pentagon $ABCDE$ in which angle A and angle E are right angles. The lengths AB, DE and EA are all 6 cm, and BC and CD are both 4 cm.

Exercise 7.5

Links 7F, G

1 Marlborough is 16 miles from Swindon on a bearing of 170°.
Chippenham is 20 miles from Marlborough on a bearing of 280°.
Using a scale of 1 cm to represent 2 miles make a scale drawing to show these places.

First make a sketch.

(a) Find the bearing of Swindon from Chippenham.

(b) Work out the distance of Swindon from Chippenham.

2 Chipping Norton is 27 km from Oxford on a bearing of 320°.
Aylesbury is 30 km from Oxford on a bearing of 072°.
Using a scale of 1 cm to represent 3 km make a scale drawing to show these places.

(a) Find the distance and bearing of Chipping Norton from Aylesbury.

Abingdon is 9 km due South of Oxford.

(b) Find the distance and bearing of Aylesbury from Abingdon.

(c) Find the distance and bearing of Chipping Norton from Abingdon.

3 Blackboard, the mathematical pirate, buried his treasure according to these instructions.

> 'The treasure lies on a bearing of 150° from the tall coconut tree and a distance of 50 metres from the spiky cactus plant.'

The spiky cactus plant is 20 metres due East of the coconut tree.

(a) Make a scale drawing to show the position of the treasure using a scale of 1 cm to represent 5 metres.

Captain Whiteboard finds the treasure and re-buries it at a point that is equidistant from the tall coconut tree and the spiky cactus plant and 25 metres from both.

(b) Show the new position of the treasure on your scale drawing.

4 Construct an accurate drawing of an equilateral triangle with sides of 8 cm.
Use compass constructions to bisect each of the angles.

5 *AB* is a line 12.5 cm long.
C is the point which is 10 cm from *A* and 6 cm from *B*.

$$\underset{\times}{C}$$

_____
A   *B*

Construct the perpendicular from *C* to *AB*.

6 Construct a triangle with sides 12 cm, 10 cm and 8 cm.
From each vertex construct the perpendicular to the opposite side.

Exercise 7.6

1 Construct the locus of the following points:

(a) 3 cm from the point Q

(b) equidistant from X and Y where $XY = 5$ cm

(c) equidistant from the lines AB and BC where angle $ABC = 60°$

(d) 2 cm from the straight line PQ where $PQ = 7.5$ cm

> **Remember:** A locus is a set of points that obeys a given rule.

2 A goat is tethered by a 10-metre-long chain in the middle of a large field. Draw, using a scale of 1 cm to represent 4 metres, the locus of the area that the goat can graze in if the chain is attached

(a) to a tree

(b) to a bar that is 20 metres long.

3 Ermintrude the cow is attached by a 15-metre-long chain to a bar that runs along the whole length of the long side of a barn that is located in the middle of a large field, as shown in the diagram. Using a scale of 1 cm to represent 2 metres draw the locus of the area in which she can graze.

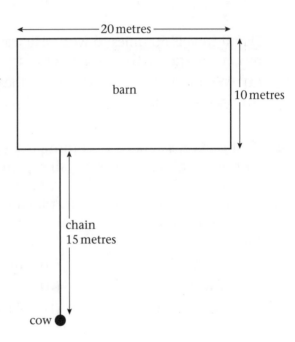

8 Fractions

1 Copy each of these shapes. In each case write down the fraction of the shape that has been shaded.

(a)   **(b)**   **(c)**   **(d)**

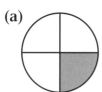 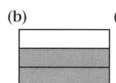 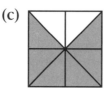 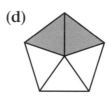

2 Make four copies of this shape. Shade them to show each of the fractions.

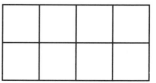

   (a) $\frac{1}{4}$ (b) $\frac{3}{4}$ (c) $\frac{5}{8}$ (d) $\frac{6}{8}$

3 David travels 60 miles to work. He travels the first 20 miles by car. He travels the remainder of the distance by rail.

   (a) What fraction of the journey does David travel by car?

   (b) What fraction of the journey does David travel by rail?

4 Yvonne keeps a diary. According to her diary, during the 24-hour period last Tuesday, she spent 8 hours at school, 1 hour travelling, 3 hours at work, 6 hours asleep, 2 hours doing homework and 4 hours playing games at home. Write down the fraction of this 24-hour period Yvonne spent

   (a) at school   **(b)** travelling

   (c) at work   **(d)** asleep

   (e) doing homework   **(f)** playing games at home.

5 Change each of these improper fractions to a mixed number.

   (a) $\frac{7}{6}$ **(b)** $\frac{17}{6}$ **(c)** $\frac{22}{7}$ **(d)** $\frac{43}{10}$ **(e)** $\frac{28}{24}$ **(f)** $\frac{43}{4}$

   (g) $\frac{56}{5}$ **(h)** $\frac{107}{10}$ **(i)** $\frac{36}{32}$ **(j)** $\frac{27}{4}$ **(k)** $\frac{135}{12}$

> $\frac{7}{6} = \frac{6}{6} + \frac{1}{6} = 1\frac{1}{6}$

6 Change each of these mixed numbers to an improper fraction.

   (a) $2\frac{1}{4}$ **(b)** $3\frac{1}{2}$ **(c)** $4\frac{2}{5}$ **(d)** $5\frac{3}{8}$ **(e)** $8\frac{1}{5}$ **(f)** $10\frac{2}{3}$

   (g) $3\frac{1}{4}$ **(h)** $5\frac{7}{8}$ **(i)** $4\frac{3}{5}$ **(j)** $12\frac{2}{3}$ **(k)** $25\frac{3}{4}$

> $2\frac{1}{4} = \frac{8}{4} + \frac{1}{4} = \frac{9}{4}$

7 Wasim bought a shirt and a tie. The cost of the shirt
was £18. The total cost of the shirt and the tie was £22.

   (a) Work out the cost of the tie as a fraction of
      (i) the cost of the shirt
      (ii) the total cost of the shirt and tie.

   (b) Work out the cost of the shirt as an improper fraction
      of the cost of the tie.

   (c) Change your answer to **(b)** into a mixed number.

Exercise 8.2   **Links 8D–F**

1 Find

> **Remember:** 'of' means '×'.

   (a) $\frac{1}{2}$ of 80   **(b)** $\frac{2}{3}$ of 60   **(c)** $\frac{4}{5}$ of 100
   (d) $\frac{3}{7}$ of 140   **(e)** $\frac{3}{4}$ of £28.40

2 A company makes $\frac{3}{8}$ of its 240 employees redundant.
How many employees are made redundant?

3 Joan's gross pay is £1200 per month. Her stoppages are $\frac{2}{5}$ of
her gross pay. Work out Joan's stoppages per month.

4 The cost of a new motor cycle is £3600. Andrew paid a
deposit which is $\frac{5}{8}$ of the cost. Work out the deposit paid by
Andrew.

5 Jaqui travelled 48 miles to see her sister. She travelled the
first 18 miles by car and the remainder by rail. Work out,
in their simplest form

   (a) the fraction of the journey Jaqui travelled by car

   (b) the fraction $\dfrac{\text{distance travelled by car}}{\text{distance travelled by rail}}$.

6 Simplify each of these fractions by finding common
factors.

> Cancel the top and the
> bottom of each fraction
> by the common factor.

   (a) $\frac{6}{8}$ **(b)** $\frac{8}{10}$ **(c)** $\frac{12}{24}$ **(d)** $\frac{24}{36}$ **(e)** $\frac{75}{100}$ **(f)** $\frac{54}{90}$

7 Last season Lucea hockey club played 40 matches.
They won 32, lost 6 and drew 2. Writing each
fraction in its simplest form, find the fraction of
the 40 matches that Lucea hockey club

   (a) won

   (b) lost

   (c) drew.

8 Copy this shape. Find two equivalent fractions to describe how much of the shape has been shaded.

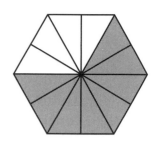

9 Make these fractions equivalent.

(a) $\frac{3}{4} = \frac{9}{?}$ (b) $\frac{16}{20} = \frac{}{5}$ (c) $\frac{1}{3} = \frac{}{24}$ (d) $\frac{35}{100} = \frac{7}{?}$

10 Write down three fractions equivalent to each of these.

(a) $\frac{4}{5}$ (b) $\frac{2}{7}$ (c) $\frac{18}{24}$ (d) $\frac{27}{36}$ (e) $\frac{30}{100}$

Exercise 8.3   Links 8G–I

1 Which is the larger

(a) $\frac{2}{3}$ or $\frac{3}{5}$ (b) $\frac{3}{4}$ or $\frac{4}{5}$ (c) $\frac{7}{10}$ or $\frac{17}{20}$?

2 Put these fractions in order, starting with the smallest.

$\frac{3}{10}$ $\frac{1}{3}$ $\frac{2}{7}$ $\frac{4}{15}$ $\frac{29}{100}$

> Write each fraction as equivalent fractions.

3 Work out

(a) $\frac{3}{7} + \frac{2}{7}$ (b) $\frac{4}{13} + \frac{8}{13}$ (c) $\frac{1}{8} + \frac{5}{8}$ (d) $\frac{2}{5} + \frac{3}{5}$

(e) $\frac{3}{8} + \frac{1}{2}$ (f) $\frac{1}{3} + \frac{1}{6}$ (g) $\frac{5}{12} + \frac{3}{4}$ (h) $\frac{3}{4} + \frac{9}{20}$

(i) $\frac{3}{5} + \frac{3}{10}$ (j) $\frac{2}{15} + \frac{1}{30}$ (k) $\frac{5}{12} + \frac{1}{3}$ (l) $\frac{2}{5} + \frac{7}{20}$

> **Remember:** It is easy to add fractions when the denominator (bottom) is the same.

4 Work out

(a) $\frac{1}{6} + \frac{3}{8}$ (b) $\frac{3}{5} + \frac{3}{8}$ (c) $\frac{2}{9} + \frac{1}{12}$ (d) $\frac{3}{10} + \frac{4}{15}$

(e) $\frac{1}{3} + \frac{1}{5}$ (f) $\frac{1}{7} + \frac{1}{6}$ (g) $\frac{2}{5} + \frac{3}{7}$ (h) $\frac{3}{5} + \frac{1}{8}$

(i) $\frac{4}{5} + \frac{1}{3}$ (j) $\frac{7}{12} + \frac{4}{9}$ (k) $3\frac{1}{2} + 2\frac{1}{4}$ (l) $4\frac{1}{3} + 3\frac{1}{4}$

> It is easier to add the whole numbers and then add the fractions.

5 George won some money in the Lottery.
He gave $\frac{2}{5}$ of this money to his wife.
He gave $\frac{1}{3}$ of this money to his daughter.
What fraction of the money did his wife and daughter receive altogether?

6 Work out

(a) $\frac{7}{8} - \frac{1}{8}$ (b) $\frac{5}{8} - \frac{1}{2}$ (c) $\frac{1}{3} - \frac{1}{5}$ (d) $\frac{1}{2} - \frac{1}{7}$

(e) $\frac{3}{8} - \frac{1}{3}$ (f) $\frac{4}{5} - \frac{1}{7}$ (g) $\frac{1}{3} - \frac{1}{8}$ (h) $\frac{2}{3} - \frac{2}{5}$

(i) $\frac{5}{7} - \frac{3}{10}$ (j) $2\frac{1}{2} - \frac{1}{4}$ (k) $3\frac{1}{4} - \frac{1}{2}$ (l) $5 - 1\frac{3}{4}$

(m) $4\frac{1}{4} - 1\frac{3}{8}$ (n) $3\frac{1}{3} - 1\frac{1}{4}$ (o) $5\frac{2}{3} - 3\frac{7}{8}$ (p) $4 - 2\frac{3}{8}$

(q) $4\frac{2}{3} - 1\frac{1}{2}$ (r) $17\frac{1}{4} - 3\frac{2}{5}$

> **Remember:** It is easy to subtract fractions when the denominator (bottom) is the same.

7 The farmer had $12\frac{1}{2}$ acres of land.
He sold $10\frac{5}{8}$ acres to a builder.
How much land did the farmer
have left after the sale?

Exercise 8.4
Links 8J, K

1 Work out

Remember:
Multiply the numerators
(top). Multiply the
denominators (bottom).

(a) $\frac{1}{2} \times \frac{1}{4}$ (b) $\frac{5}{8} \times \frac{1}{4}$ (c) $\frac{2}{3} \times \frac{4}{5}$

(d) $\frac{3}{5} \times \frac{7}{8}$ (e) $\frac{2}{5} \times \frac{3}{4}$ (f) $\frac{3}{7} \times \frac{7}{12}$

(g) $\frac{1}{4} \times 20$ (h) $\frac{2}{5} \times 8$ (i) $\frac{3}{14} \times 280$

(j) $57 \times \frac{2}{3}$ (k) $24 \times \frac{3}{8}$ (l) $15 \times \frac{7}{8}$

(m) $2\frac{1}{4} \times 3\frac{1}{2}$ (n) $4\frac{3}{4} \times 5\frac{1}{2}$ (o) $1\frac{1}{3} \times \frac{1}{4}$

(p) $3\frac{2}{5} \times 4\frac{1}{7}$ (q) $6\frac{1}{2} \times 4\frac{1}{4}$ (r) $2\frac{1}{5} \times 1\frac{7}{8}$

Change mixed numbers
to improper fractions
first.

2 A rectangular carpet measures $3\frac{1}{4}$ metres by $4\frac{1}{8}$ metres.
Work out the area of the carpet.

3 The capacity of a bucket is $6\frac{1}{3}$ litres.
$3\frac{1}{2}$ buckets full of water are poured into an large empty
container. How much water does this container now hold?

4 A building brick weighs $1\frac{1}{3}$ kg.
Work out the weight of 6 bricks.

5 Work out

Remember: Turn the
fraction you are dividing
by upside down, and
multiply.

(a) $\frac{1}{2} \div \frac{1}{4}$ (b) $\frac{7}{8} \div \frac{1}{4}$ (c) $2\frac{3}{4} \div 1\frac{1}{2}$

(d) $\frac{7}{8} \div 4$ (e) $\frac{2}{7} \div 2$ (f) $\frac{5}{7} \div 3$

(g) $12 \div \frac{1}{4}$ (h) $7 \div \frac{1}{3}$ (i) $\frac{3}{7} \div \frac{1}{4}$

(j) $\frac{4}{5} \div \frac{5}{9}$ (k) $2\frac{1}{2} \div 3\frac{1}{4}$ (l) $4\frac{2}{7} \div 3\frac{5}{8}$

Turn mixed numbers into
improper fractions.

6 A bag of flour weighs 3 kg. It takes $\frac{3}{5}$ kg of flour to make a
loaf of bread.
How many loaves of bread can 2 bags make?

Exercise 8.5
Link 8L

1 A salt cellar holds $2\frac{1}{2}$ ounces of salt. A tub of salt holds
20 ounces. How many salt cellars can be filled from the tub?

2 Find the difference between $\frac{4}{5}$ of 45 miles and $\frac{7}{8}$ of 48 miles.

3 On a coach 25 of the passengers are women, 15 are men and 5 are children. What fraction of the total number of people are women?

4 A metal bar is 120 cm long. When heated it expands by $\frac{1}{8}$ of the original length. What is its length when heated?

5 Roger had 96 sheep last year. His flock increased by $\frac{3}{8}$ this year.
How many sheep does Roger have now?

6 A sack of potatoes weighs 56 kg. More potatoes are added and the weight of the sack of potatoes increases by $\frac{3}{7}$.
What is the new weight of the sack of potatoes?

Exercise 8.6   Links 8M–O

1 Change these fractions to decimals. Show your working.
  (a) $\frac{1}{4}$ (b) $\frac{2}{5}$ (c) $\frac{9}{10}$ (d) $\frac{7}{100}$ (e) $\frac{3}{20}$
  (f) $\frac{3}{8}$ (g) $\frac{1}{3}$ (h) $\frac{2}{7}$ (i) $\frac{37}{50}$ (j) $\frac{54}{90}$
  (k) $\frac{17}{200}$ (l) $\frac{21}{40}$ (m) $\frac{18}{35}$ (n) $\frac{637}{1000}$ (o) $\frac{13}{2000}$

2 Change these decimals to fractions.
  (a) 0.7   (b) 0.41   (c) 0.253   (d) 0.8
  (e) 0.35   (f) 0.173   (g) 0.99   (h) 0.09
  (i) 0.011   (j) 0.0041   (k) 0.0909   (l) 0.0035

3 Write these fractions as decimals.
  (a) $2\frac{5}{16}$ (b) $5\frac{3}{5}$ (c) $3\frac{9}{25}$ (d) $4\frac{7}{26}$

4 Write these decimals as fractions.
  (a) 2.125   (b) 10.7142   (c) 19.173   (d) 3.013

5 Write these fractions as recurring decimals.
  Write your answers
  (i) as shown on the calculator display
  (ii) using recurring decimal notation.
  (a) $\frac{2}{3}$ (b) $1\frac{7}{9}$ (c) $4\frac{3}{11}$ (d) $2\frac{5}{6}$
  (e) $3\frac{2}{9}$ (f) $\frac{56}{63}$ (g) $\frac{7}{13}$ (h) $4\frac{13}{44}$

6 Find the reciprocals of these numbers
  (i) using a calculator   **(ii)** without using a calculator.
  (a) 3   (b) 2   (c) $\frac{1}{3}$ (d) $\frac{2}{5}$
  (e) 1000   (f) 0.001   (g) $\frac{3}{40}$ (h) $\frac{8}{5}$
  (i) 0.110   (j) 150   (k) $\frac{11}{6}$ (l) 0.375

9 Estimating and using measures

Exercise 9.1 Link 9A

Look at these pictures, then write down an estimate for each of the following.

1 The height of the bus.

2 The width of the bus.

3 The height of the litter bin.

4 The height of the woman.

5 The height of the lamp-post.

6 The diameter of the bicycle wheel.

7 The length of the car.

8 The height of the car.

9 The diameter of the wheel.

10 The height of the window from the ground.

11 The width of the window.

12 The width of the car.

Exercise 9.2 Link 9B

Write down an estimate of the amount of liquid in each of these containers. Give your answers in metric and Imperial units.

1 The coffee cup.

2 The coffee maker.

3 The mug of tea.

4 The glass of milk with a straw in it.

5 The wine glass.

6 The bottle of champagne.

7 The jug when it is full.

8 The glass next to the jug.

Exercise 9.3

Link 9C

Write down an estimate of the weight of the following fruits.
Give your answers in metric and Imperial units.

1   The 4 bananas.

2   The 3 kiwi fruit.

3   The raspberries.

4   The 4 pears.

5   The pineapple.

6   The bowl of blackberries.

7   A large bag of potatoes.

8   The cabbage.

9   The bag of onions.

10   The box of apples.

11   The basket of potatoes.

12   The melon.

Exercise 9.4

Link 9D

For each of these statements say whether the measurements are
sensible or not. If the statement is not sensible then give a
reasonable estimate for the measurement.

1   My mother is 1.60 m tall.

2   My 15-year-old brother is 2 cm tall.

3   The front door of my house is 20 m high.

4   My bedroom measures 30 m by 20 m.

5   An apple weighs 2 kg.

6   A can of cola holds about 400 m*l* of liquid.

7   This page is about 27 mm long.

8   A loaf of bread weighs about 800 g.

9   A box of chocolates weighs about 500 kg.

10 This book weighs about 10 kg.

11 The petrol tank in a small car holds about 50 litres.

12 A large bottle of cola holds about 3 m*l* of liquid.

13 A tea cup, when full, holds about 250 m*l*.

14 My car is capable of travelling at 500 km per hour.

15 A high-speed train can travel at 100 miles an hour.

Exercise 9.5   Link 9E

Copy and complete this table with appropriate units for each measurement. Give both metric and Imperial units of measurement.

	Metric	Imperial
1 Your height.		
2 The length of your desk.		
3 The weight of a packet of chocolates.		
4 The weight of a large bag of potatoes.		
5 Your weight.		
6 The amount of water in a water barrel.		
7 The amount of liquid in a petrol tanker.		
8 The amount of wine in a wine bottle.		
9 The time it takes to walk one mile.		
10 The time it takes to run 100 metres.		
11 The time it takes to drive 500 miles.		
12 The thickness of this book.		
13 The weight of a tube of sweets.		
14 The time it takes to travel from Earth to Jupiter.		
15 The diameter of a football.		

Exercise 9.6   Links 9F–H

1 Write down in words the time shown on these clocks.

(a) 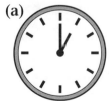 **(b)**  **(c)**  **(d)**

2 Bill arrives at the bus station and looks at his watch.
The next bus is due to arrive at 5:30 pm.
How long does Bill have to wait until the bus should
arrive?

3 Draw four clock faces and mark these times on them.

(a) quarter past 3   (b) quarter to 9

(c) twenty past 4   (d) ten to 5

4 Draw four digital watches and mark these times on them.

(a) quarter to 6   (b) half past 2

(c) twenty-five to 7   (d) five past 4

5 Seamus arrives at the train station and looks at the time
shown on the station clock. His train is due to arrive at
10:35 am.

(a) How long does he have to wait until the train should
arrive?

The train arrives 6 minutes late.

(b) How long did Seamus have to wait for his train?

6 Change these times from 12-hour clock times to 24-hour
clock times.

Remember:
3 am is 03:00 and
3 pm is 15:00

(a) 9:00 am   (b) 9:00 pm

(c) 6:30 am   (d) 6:45 pm

(e) 1 am   (f) 2 pm

(g) a quarter to eight in the morning

(h) a quarter past seven in the evening

7 Change these times from 24-hour clock times to 12-hour
clock times (am or pm).

(a) 07:00   (b) 17:00

(c) 15:30   (d) 08:50

(e) 18:50   (f) 07:30

(g) 00:10   (h) 23:45

Exercise 9.7

1 Write down the readings on these scales.

(a)

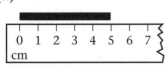

(b)

(c)

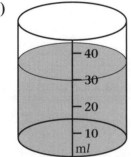

(d)

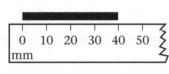

2 Write down the readings on these scales.

(a)

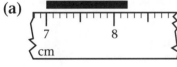

(b)

(c)

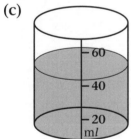

(d)

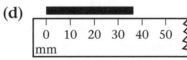

3 Write down the cost of posting these letters by
(i) first class   **(ii)** second class.

(a)

First class	30p	46p	64p	79p	94p	£1.07	£1.21	£1.40
		100g ↑		200g		300g		400g
Second class	21p	35p	47p	58p	71p	85p	94p	£1.14

(b)

First class	30p	46p	64p	79p	94p	£1.07	£1.21	£1.40
		100g		200g		300g ↑		400g
Second class	21p	35p	47p	58p	71p	85p	94p	£1.14

4 Measure and write down the lengths of these lines in centimetres.

(a) ————————————————————

(b) —————————————————————

(c) ———————————————

(d) ————————

(e) —————————————————————

5 Draw and label lines of length

    (a) 6 cm   (b) 5.5 cm   (c) 3.8 cm   (d) 10.3 cm

    (e) 23 mm   (f) 55 mm   (g) 30 mm   (h) 112 mm

6 Write down the units you would use to measure

    (a) the length of the double decker bus

    (b) the height of the double decker bus

    (c) the weight of the double decker bus

    (d) the capacity of the fuel tank of the double decker bus.

7 Write down the units you would use to measure

    (a) the distance of your school from your home

    (b) the time it takes you to travel from home to school.

8 Draw two lines that are 8 cm long.

    (a) Mark a point that is half way along the first line.

    (b) Mark a point that is a quarter of the way along the second line.

9 Draw two lines that are 9 cm long.

    (a) Mark a point that is one third of the way along the first line.

    (b) Mark a point that is two thirds of the way along the second line.

10 Here is a diagram for converting between litres and pints.

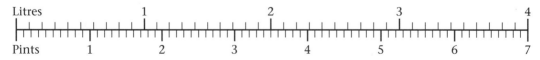

Use the diagram to convert

    (a) 2 litres to pints   (b) 3 litres to pints

    (c) 5 pints to litres   (d) 2 pints to litres.

10 Collecting and recording data

1 Here are some questions that are not suitable for a questionnaire.
For each one, say what was wrong with it and write a more suitable question.

(a) Terry wants to carry out a survey on the number of hours that people watch television. He asks:
'Do you watch television?'
a lot ☐   a little ☐   never ☐

(b) Fez is carrying out a survey on crime in his community. He asks:
'Are the police OK?'

(c) Anthea was carrying out a survey on favourite breakfast cereals. She asks:
'Do you like cereals?'

(d) Stina wants to find out what people think about paying for dental care. She asks:
'Do you agree that all dental care should be free?'

(e) Michael wants to find out what type of music people like to listen to. He asks:
'Do you like pop music or classical music?'

(f) Sethina wants to find about the types of things people recycle. She asks:
'What do you recycle?'
Paper ☐   Glass ☐

(g) David carries out a survey on which party people voted for in the general election. He asks:
'You did vote for the Monster Raving Loony Party, didn't you?'
Yes ☐   No ☐

(h) Kate asks her friends about their mock exam results. She asks:
'How well did you do in your mock exams?'
badly ☐   well ☐   OK ☐

BALLOT	
Place your **X** in one box only	
Labour	☐
Liberal Democrats	☐
Conservative	☐
Monster Raving Loony	☐

2 Draw up a questionnaire to find out about people's hobbies.

3 Design a questionnaire to find out what people are looking for when they choose a car.

Exercise 10.2

Links 10C–E

1 In each of these cases choose the most appropriate group of people you would use to collect your data.

(a) Where people eat their lunch.
   (i) by asking the first 10 people in the lunch queue
   (ii) by asking every third person in your tutor group
   (iii) by asking the first 10 people going home for lunch

(b) What you had for breakfast.
   (i) by asking a random sample of people in your school
   (ii) by asking all those people who were late for school
   (iii) by asking all those people on the school bus

(c) How you travelled to school.
   (i) by asking a random selection of 30 people in the canteen
   (ii) by asking all those people who were late for school
   (iii) by asking all those people on the school bus

2 A new supermarket is to be built. The company carry out a survey to find people views on where it should be built. There are three sites where it could be built. One site is in the middle of town near three other supermarkets, one site is on the edge of town with its own free car park and the other site is next to the football ground.

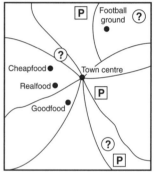

(?) Proposed new supermarket

(a) Suggest a suitable sample and where it should be collected from.

(b) Design a suitable questionnaire that the company could use to collect their data.

3 Design a suitable data collection sheet to find out about the lengths of the songs in your music collection. Use your data collection sheet to check how easy it is to use.

4 Design a suitable data collection sheet to find out about the numbers and types of coins that people are carrying with them. Use your data collection sheet to check how easy it is to use.

5 Carry out the following experiments and record your results in a tally table. You should repeat each experiment 50 times.

(a) Roll two dice, add the scores on each one and record the result.

(b) Roll two dice, subtract the scores on each one and record the result.

(c) Throw two coins in the air and record the number of times they come down with 2 heads or 2 tails or a head and a tail.

Exercise 10.3

Links 10F, G

1   Here is some data about the students in Gareth's maths class.

Name	Tutor group	Test mark	Estimated GCSE grade	Number of detentions	Number of merits	Parents seen
Gareth	11A	50%	E	0	5	Yes
Naomi	11A	46%	E	2	2	Yes
Nilmini	11C	67%	D	1	1	No
Moshe	11D	72%	D	0	6	Yes
Eira	11A	35%	F	4	2	No
Flora	11C	42%	E	2	3	No
Iain	11C	55%	E	4	0	Yes

Use the data to answer the following questions.

(a)   Which students scored more marks in the test than Gareth?

(b)   Which tutor group was only one person in?

(c)   Write down the names in the order of the number of merits received.

(d)   How many students' parents had Gareth's teacher seen?

(e)   Which student had the lowest estimated GCSE grade?

(f)   Write down the names in the order of their test marks.

(g)   Which students had had no detentions?

2   Here is some data about four used cars available from the internet.

Make	Model	Year	Size of engine	Number of miles	Number of owners	Price
Volvo	850 SE	1997	2500	12 000	2	£5500
Saab	9000i	1996	2000	25 000	3	£3500
Honda	Civic	1996	2000	30 000	2	£2500
Rover	420i	1995	1600	45 000	1	£1995

(a)   Which is the cheapest car?

(b)   Which car has the biggest engine?

(c)   Write down the cars in order of price.

(d)   Which car is the oldest?

(e)   Which car has had the most owners?

3

UK POPULATION, 1901 – 2001

Population in thousands

	UNITED KINGDOM	ENGLAND	WALES	SCOTLAND	NORTHERN IRELAND
1901	38,237	30,515	2,013	4,472	1,237
1911	42,082	33,649	2,421	4,761	1,251
1921	44,027	35,231	2,656	4,882	1,258
1931	46,038	37,359	2,593	4,843	1,243
1951	50,225	41,159	2,599	5,096	1,371
1961	52,807	43,561	2,635	5,184	1,427
1971	55,928	46,412	2,740	5,236	1,540
1981	56,352	46,821	2,813	5,180	1,538
1993	58,191	48,533	2,906	5,120	1,632
2001	59,800	50,023	2,966	5,143	1,667

Source: *Annual Abstract of Statistics*, 2005

(a) What was the population of Wales in 1971?

(b) How much did the population of the United Kingdom rise between 1961 and 1971?

(c) In what year was the population of Scotland the highest?

(d) Between which years did the population of Northern Ireland fall and then rise again?

(e) Between which years did the population of England rise the least?

4 Here is some data about British Prime Ministers from 1957 to 1997.

Name	Party	Years in office	Born	Age at taking office	Time in office
Callaghan, James	Labour	1976–79	1912	64	3 years, 29 days
Douglas-Home, Alec	Conservative	1963–64	1903	60	362 days
Heath, Edward	Conservative	1970–74	1916	53	3 years, 259 days
Macmillan, Harold	Conservative	1957–63	1894	62	6 years, 281 days
Major, John	Conservative	1990–97	1943	47	6 years, 154 days
Thatcher, Margaret	Conservative	1979–90	1925	53	11 years, 209 days
Wilson, Harold	Labour	1964–70 and 1974–76	1916	48	7 years, 279 days

Use the data to answer the following questions.

(a) Which Prime Minister was the youngest at taking office?

(b) Which Prime Minister had the shortest time in office?

(c) When was Edward Heath born?

(d) Write down all the Labour Prime Ministers.

(e) Which Prime Minister took office in 1957? How long was he in office?

(f) Write down the names in order of when they took office.

Exercise 10.4   Link 10H

1 The mileage chart shows the distances between some European cities. Use the information to answer the following.

London					
215	Paris				
908	689	Rome			
774	641	844	Madrid		
791	648	477	1122	Vienna	
468	245	444	627	507	Geneva

    (a) What is the distance between London and Rome?

    (b) Which two cities are closest?

    (c) Which city is nearest to Madrid?

    (d) Work out the total distance of a round-trip that goes from London to Vienna, then from Vienna to Geneva, and finally from Geneva back to London.

    (e) Work out the difference between the distance from Paris to Madrid and the distance from Paris to Rome.

2 Here is a map of the roads between four towns.

Use the distances to draw up a mileage chart for these towns.

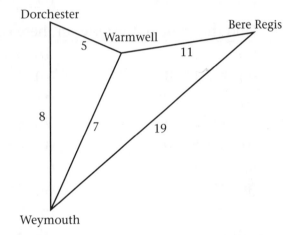

3 Here are the points *A*, *B*, *C* and *D*.

Measure the distance between each pair of points.
Use your results to draw up a distance chart for these points.

Measure the distances to the nearest mm

4 Find out the distances between any places of local interest close to your school.

Present the information in a distance chart.

11 Linear equations

1 Find the value of the letter in these equations.

(a) $a + 3 = 7$ (b) $b + 2 = 6$

(c) $c + 4 = 5$ (d) $w - 2 = 6$

(e) $x + 5 = 9$ (f) $y - 3 = 5$

(g) $d + 6 = 11$ (h) $m - 6 = 7$

(i) $10 + x = 19$ (j) $15 = y + 3$

(k) $x - 6 = 0$ (l) $8 - x = 5$

(m) $p + 9 = 16$ (n) $r - 3 = 7$

(o) $a + 4 = 9$ (p) $b - 3 = 4$

(q) $d + 1 = 14$ (r) $10 - f = 6$

> **Remember:** Always do the same thing to both sides of the equation, e.g. if you take 3 away from the left-hand side, you must also take 3 away from the right-hand side.

2 Find the value of the letter in these equations.

(a) $x + 5 = 9$ (b) $a + 2 = 7$

(c) $c + 1 = 8$ (d) $t - 4 = 3$

(e) $m - 5 = 8$ (f) $n - 8 = 9$

(g) $y + 5 = 5$ (h) $s + 5 = 7$

(i) $x - 9 = 0$ (j) $d + 4 = 15$

(k) $f - 3 = 11$ (l) $4 - y = 2$

(m) $10 = x + 2$ (n) $15 - x = 7$

(o) $x - 18 = 25$ (p) $8 = y + 8$

(q) $t - 7 = 10$ (r) $8 = r - 3$

(s) $15 = 8 + x$ (t) $22 = 12 + x$

(u) $11 = x - 11$

> You may find the balancing method helps you.

Find the value of the letter in each of these equations.

1 $2x = 4$ **2** $3y = 9$ **3** $2a = 6$

4 $4y = 12$ **5** $5c = 20$ **6** $3y = 18$

7 $9t = 18$ **8** $4w = 24$ **9** $7x = 35$

10 $3x = 10$ **11** $\dfrac{x}{2} = 4$ **12** $\dfrac{y}{3} = 12$

13 $\dfrac{a}{7} = 3$ **14** $\dfrac{b}{4} = 5$ **15** $\dfrac{c}{7} = 1$

> Look for the opposite (or inverse) process. The inverse of ÷3 is ×3. Multiply each side of the equation by 3.

16 $\dfrac{t}{4} = 6$ **17** $\dfrac{a}{11} = 5$ **18** $\dfrac{y}{2} = 13$

19 $\dfrac{f}{4} = 4$ **20** $\dfrac{m}{9} = 3$ **21** $x + 9 = 18$

22 $b - 4 = 13$ **23** $15 - x = 1$ **24** $7g = 42$

25 $\dfrac{a}{5} = 5$ **26** $x + 11 = 25$ **27** $y - 7 = 18$

28 $r - 7 = 0$ **29** $3 - x = 11$ **30** $15 = y + 12$

31 $9x = 72$ **32** $15 = \dfrac{a}{3}$ **33** $45 = 5x$

34 $11 + t = 19$ **35** $11 - s = 16$ **36** $6p = 24$

Exercise 11.3   Links 11G–J

Find the value of the letter in each of these equations.

1 $2x + 1 = 5$ **2** $2x - 3 = 5$ **3** $2x + 5 = 7$

4 $3y + 4 = 13$ **5** $4a - 5 = 11$ **6** $5t + 1 = 26$

7 $6m - 3 = 18$ **8** $9x + 6 = 33$ **9** $4a + 1 = 11$

10 $2b + 5 = 1$ **11** $3d + 4 = 1$ **12** $4x + 9 = 5$

13 $5t + 11 = 7$ **14** $2x + 7 = 3$ **15** $3x - 2 = -14$

16 $3y + 7 = 2$ **17** $4x - 10 = -10$ **18** $2a + 15 = 3$

19 $6s + 9 = 30$ **20** $4 + 2x = 8$ **21** $10 + 2y = 6$

22 $12 = 2x + 5$ **23** $6k - 9 = 15$ **24** $3x - 7 = -11$

25 $12x + 5 = 5$ **26** $9y + 9 = 0$ **27** $15a + 1 = 46$

28 $4t - 13 = 19$ **29** $3g + 11 = 23$ **30** $5x + 7 = 31$

31 $\dfrac{x}{2} + 3 = 5$ **32** $\dfrac{y}{3} - 7 = 1$ **33** $\dfrac{r}{3} - 1 = -4$

34 $\dfrac{x}{7} - 1 = 2$ **35** $\dfrac{a}{3} + 2 = 5$ **36** $-\dfrac{c}{4} + 1 = 3$

Exercise 11.4   Links 11K, L

Solve these equations.

1 $2(x + 3) = 10$ **2** $2(y - 3) = 6$ **3** $3(a + 4) = 21$

4 $4(d - 4) = 12$ **5** $3(t + 5) = 24$ **6** $2(w - 5) = 8$

7 $7(x + 1) = 21$ **8** $4(x + 5) = 12$ **9** $5(d + 1) = 0$

10 $3(2d + 1) = 9$ **11** $2(3x - 2) = 14$ **12** $2(5y - 3) = 14$

13 $2(5p - 2) = 46$ **14** $4(2x - 1) = 3$ **15** $3(4y - 3) = 7$

Remember: 'Solve' means find the value of the letter.

Expand the brackets first.

Exercise 11.5

Links 11M, N

Solve these equations.

1 $2x + 3 = x + 7$

2 $5x - 3 = 4x + 1$

3 $5y + 7 = 3y - 5$

4 $4y - 3 = 3y + 1$

5 $12t - 8 = 7t + 7$

6 $3x - 8 = x + 4$

7 $5a + 3 = 3a - 4$

8 $2b + 7 = 8b - 3$

9 $11c - 9 = 4c + 3$

10 $5x - 2 = 3x - 7$

11 $8y - 9 = 5y + 6$

12 $4s - 3 = 7s - 15$

13 $3a - 2 = a + 4$

14 $3x + 5 = 5x + 9$

15 $5x - 9 = 2x + 1$

16 $4y - 11 = 7y + 4$

17 $3s + 8 = 7s - 9$

18 $8g + 1 = 15g - 6$

19 $9y + 15 = 7y - 3$

20 $4x + 5 = 3x - 4$

First try to get all the letters on one side of the equation only.

12 Sorting and presenting data

1 Yvonne and Gulzar conducted a survey into the colours of cars. They recorded the colour of 50 cars which passed the school gates one morning. The results of the survey are given below.

blue	white	red	red	grey
green	white	black	yellow	red
blue	grey	red	green	red
white	red	blue	yellow	blue
red	white	grey	red	red
black	red	red	green	blue
red	blue	brown	red	grey
silver	red	silver	red	white
black	grey	grey	white	black
yellow	green	red	blue	red

(a) Draw up a tally chart and frequency table for these colours.

(b) What was the most popular colour for a car?

(c) Draw a bar chart to show the results of the survey.

> **Remember:** '5' is represented by |||| in a tally chart.

> **Remember:** Leave gaps between the bars on your bar chart.

2 As part of a coursework, Kevin asked 40 people which day of the week they were most likely to go to the shops to do their main shopping for the week. The results of this survey are shown below.

Sat	Tues	Sat	Thurs	Fri	Thurs	Wed	Sat
Sun	Wed	Thurs	Fri	Tues	Sat	Mon	Thurs
Thurs	Sat	Fri	Thurs	Thurs	Sat	Thurs	Wed
Sat	Thurs	Thurs	Sat	Thurs	Thurs	Thurs	Fri
Fri	Sat	Thurs	Tues	Sun	Fri	Sat	Thurs

For this data

(a) draw up a tally chart and frequency table

(b) draw a bar chart.

(c) Which is
  (i) the most popular day for people to do their main shopping
  (ii) the least popular day for people to do their main shopping?

3 Fifty students sat an examination in English. The numbers of spelling mistakes they made are recorded in the table below.

Spelling mistakes

6	5	2	8	10	5	9	8	1	12
10	8	7	7	13	12	2	3	8	7
7	6	15	12	10	8	5	11	15	5
8	8	2	6	7	2	8	12	14	3
4	9	8	5	4	14	12	8	3	8

For this data

(a) draw up a tally and frequency chart

(b) draw a bar chart

(c) find the mode of the number of mistakes.

> **Remember:** The mode is the value which occurs most often.

4 There are 80 members of Wellshall Theatre Society. The ages, in years, of the members are given below.

Ages in years

27	19	32	46	8	17	33	42	64	71
14	34	33	52	30	41	38	32	24	61
35	36	29	42	57	34	36	19	18	45
36	22	28	42	50	17	73	7	12	62
42	55	26	32	33	39	27	31	28	44
33	46	23	17	48	41	37	32	28	66
78	32	45	33	20	18	44	53	38	39
34	31	42	23	16	14	58	37	30	22

(a) Draw up a frequency table using class intervals of 0 to 9, 10 to 19, etc.

(b) Illustrate the data with a bar chart.

(c) Write down the modal class interval.

The society is to accept some new members.

(d) What is the minimum number of new members the society needs to accept to bring about a change in the modal class interval?

(e) Make a comment about the ages of these new members if this change occurs.

> Look at how many members there are in the second most common class.

5 Talber sells men's shirts at a market stall. Here is the size of each of the shirts he sells.

$14\frac{1}{2}$	13	$15\frac{1}{2}$	14	$16\frac{1}{2}$	15	14	$15\frac{1}{2}$	$13\frac{1}{2}$	14
14	15	16	13	$13\frac{1}{2}$	14	$14\frac{1}{2}$	16	15	17
13	14	$13\frac{1}{2}$	15	15	16	15	13	14	$14\frac{1}{2}$
15	$14\frac{1}{2}$	$15\frac{1}{2}$	16	$14\frac{1}{2}$	$13\frac{1}{2}$	$16\frac{1}{2}$	14	16	15
14	$15\frac{1}{2}$	13	14	15	$16\frac{1}{2}$	14	17	$14\frac{1}{2}$	$13\frac{1}{2}$

(a) Draw up a frequency table using class intervals 13–$13\frac{1}{2}$, 14–$14\frac{1}{2}$, ...

(b) What is the modal class?

(c) Draw a bar chart of the data.

6 Forty people took part in a clay pigeon shooting competition. Here are the points they scored.

18	24	19	3	24	11	25	10
13	26	22	20	7	20	22	25
23	5	12	27	28	21	16	14
12	25	7	25	19	17	15	8
14	25	9	16	26	21	27	25

(a) Draw up a frequency table using class intervals 1–5, 6–10, ...

(b) With the same data, draw up a frequency table using class intervals 1–3, 4–6, ...

(c) Which of your answers do you think is better? Why?

Exercise 12.2
Links 12C, D

1 The dual bar chart gives information about the life expectancy at birth in certain African countries and the UK.

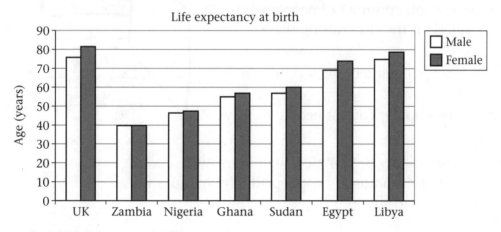

Write down three things you notice from the chart.

2 The average daily temperatures in Leeds and Athens are recorded in the table below in °C.

	Oct	Nov	Dec	Jan	Feb	Mar	Apr
Leeds	14	10	8	7	9	10	15
Athens	25	22	14	13	15	17	20

(a) Draw a dual bar chart to illustrate this data.

(b) Write down three statements about the temperatures in Leeds and Athens.

3 The pictogram shows some information about the numbers of people who played at the local squash club last week.

Sunday	
Monday	
Tuesday	
Wednesday	
Thursday	
Friday	
Saturday	

Each  represents 20 people

(a) How many played on
(i) Sunday   (ii) Monday?

On Tuesday 50 people played squash at the club.

(b) Copy and complete the pictogram.

(c) Work out the total number of people who played squash at the club during the week.

(d) Which was

   (i) the most popular day for people to play squash

   (ii) the least popular day for people to play squash?

It costs each person £1.50 to play squash at the club.

(e) What was the total amount people paid to play squash on
(i) Friday   (ii) Tuesday?

4 A travel agent kept information about the countries her customers visited for their summer holiday last year. This information was displayed in the form of a pictogram shown opposite.

(a) How many people visited
   (i) the United Kingdom
   (ii) Greece?

Last summer 300 customers visited Turkey.

(b) Copy and complete the pictogram.

(c) Which was the most popular place that the customers visited last summer?

United Kingdom	
France	
Spain	
Italy	
Greece	
Turkey	

Each ⚲ represents 50 customers

Exercise 12.3

Links 12E–G

1 Which of the following are discrete data and which are continuous data?

> **Remember:** Data which can be counted is discrete. Data which can be measured is continuous.

(a) your height

(b) your age

(c) the number of people you speak to in the next hour

(d) the amount you spend on sweets in a week

(e) your body temperature

(f) how long it takes you to go home after school

(g) the number of steps you walk in a day

2 At 8 am Jacqui started a journey in her car. The journey finished at 5 pm. She started the journey with a full tank of petrol. The line graph shows the amount of petrol in the fuel tank of Jacqui's car at various times on the journey.

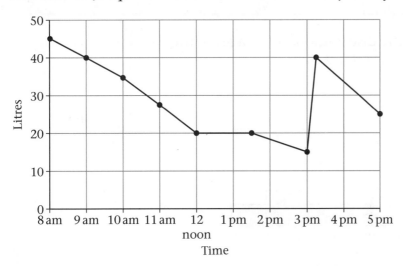

(a) How much petrol does the fuel tank hold when full?

(b) Between which times do you think Jacqui stopped for a lunch break?

(c) Describe what you think happened between 3:00 pm and 3:15 pm.

(d) How much petrol was in the tank at
   (i)   10:00 am
   (ii) 2:00 pm
   (iii) the end of the journey?

(e) At what time was the fuel tank half full?

(f) How much petrol did the car use on the whole journey?

3 This table shows the quarterly sales of pianos at Music-R-Us.

Year	2004				2005				2006			
Quarter	1	2	3	4	1	2	3	4	1	2	3	4
Sales	65	53	56	68	56	48	47	52	44	39	38	

(a) Draw a time series graph to represent this information.

(b) Comment on your graph.

(c) Make a prediction of the sales for the last quarter of 2006. Give a reason for your answer.

4 The table shows the numbers of students at Lucea High School who were late for the start of school one week.

Day	Mon	Tue	Wed	Thur	Fri
Number late	40	32	35	47	18

(a) Draw a vertical line graph to display this information.

(b) Explain why this data cannot be displayed using any other form of line graph.

5 The histogram shows the distribution of the ages of the 100 members of Aqua Swimming Club.

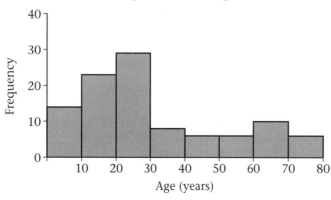

(a) Make a copy of the histogram.

(b) On the same diagram draw the frequency polygon.

The distribution of the ages of the 100 members of Lucea Swimming Club is given below.

Age	10–19	20–29	30–39	40–49	50–59	60–69	70–79
Frequency	6	15	24	28	17	8	2

(c) For this second distribution, draw on the same axes
  (i) the histogram   (ii) the frequency polygon.

(d) Use the frequency polygons to compare the two distributions of ages, writing down three observations you have found.

6 Five thousand runners took part in a marathon race.
The histogram shows part of the distribution of the times
it took the runners to complete the race.

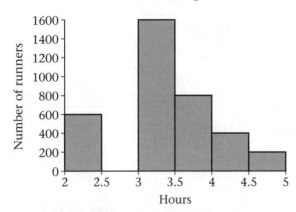

(a) How many runners completed the race in less than
2.5 hours?

1400 runners completed the race in more than 2.5 but less
than 3 hours.

(b) Copy and complete the histogram.

(c) How many people completed the race in
    (i)   less than 3 hours
    (ii) more than 3.5 hours?

7 There are 120 people living in the small hamlet of
Fellworth. The age ranges of these people are displayed in
the table below.

Age (years)	0–9	10–19	20–29	30–39	40–49	50–59	60–69	70–79
Frequency	10	15	17	28	22	15	11	2

(a) Draw a histogram for this data.

(b) On your histogram, draw a frequency polygon.

(c) What is the modal age range for the people living in
Fellworth?

(d) What percentage of the people living in Fellworth are
aged between 20 and 39 years?

13 3-D shapes

1 Copy the picture of a house and on it mark

   (a) two lines that are horizontal and parallel

   (b) two lines that are parallel and vertical

   (c) two lines which are parallel but neither vertical or horizontal

   (d) two lines which are perpendicular.

2 *ABCDEFGH* is a cube resting on a level plane.

   (a) List the edges parallel to *BC*.

   (b) List the edges parallel to *DH*.

   (c) List the edges perpendicular to *AD*.

   (d) List the edges perpendicular to *FG*.

   (e) Which face is parallel to *ADCB*?

   (f) Which faces are perpendicular to *CDHG*?

   (g) Which faces are horizontal?

   (h) Which faces are vertical?

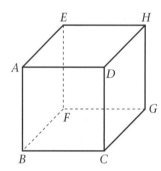

Remember: Lines and faces are **perpendicular** if the angle between them is a right angle.

3 *ABCDEF* is a triangular-based prism resting on a level plane.

   (a) Which faces are vertical?

   (b) Which faces are parallel?

   (c) List edges which are parallel to **(i)** *EF* **(ii)** *AB* **(iii)** *DF*.

   (d) List edges which are horizontal.

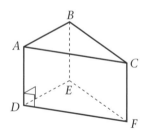

4 The boot stand *PQRSTUVW* is a prism with a trapezium cross-section. It is resting on a level floor.

   (a) Which faces are vertical?

   (b) Which faces are perpendicular to *TUVW*?

   (c) Which edges are perpendicular to *QR*?

   (d) Which edges are parallel to *PS*?

   (e) Which edges are horizontal?

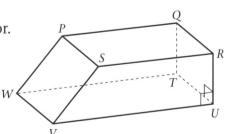

Remember: Dashed lines show edges you can't see.

Exercise 13.2

Links 13C, D

If possible, use triangular dotted paper.

1   The diagram shows two sketches
    of a 3-D letter H.

    Make sketches for 3-D letters
    (a) I
    (b) F
    (c) E
    (d) L
    (e) T

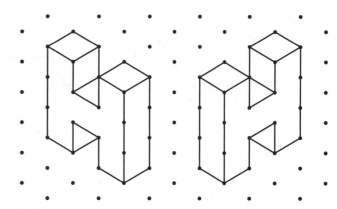

Draw the front face first.

You can also try letters A, B, C, O, P, U.

2   Sketch two hexagonal prisms with their hexagonal faces horizontal. If you are
    using triangular dotted paper do one upright and one which leans to the side.

3   Name the solid shapes shown below.

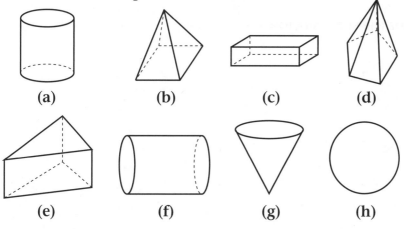

   (a)   (b)   (c)   (d)

   (e)   (f)   (g)   (h)

Exercise 13.3

Link 13E

1   Draw sketches of the shapes that can be made from these nets.

Imagine the net being folded inwards.

(a)   (b)

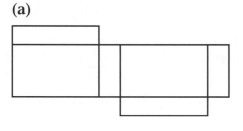

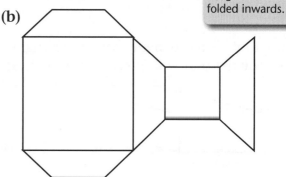

2 Which of the following are nets of a square-based
pyramid?

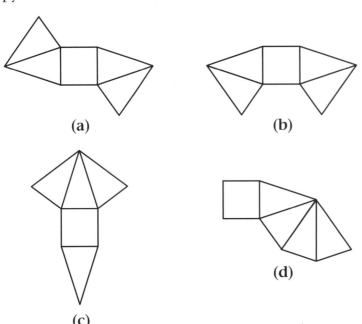

(a)   **(b)**

(c)

(d)

3 Draw accurate nets for the shapes below.

(a)

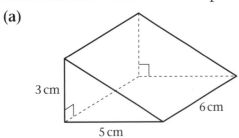

3 cm

6 cm

5 cm

(b)

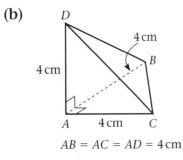

D

4 cm

4 cm   *B*

A   4 cm   *C*

AB = AC = AD = 4 cm

4 This cuboid is to be made from a rectangle of card.
Draw the net that will fit on to the smallest sheet of card.

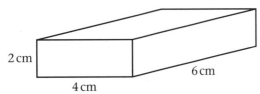

2 cm

4 cm

6 cm

Exercise 13.4

1 Sketch the plan and elevation of these solid shapes.

(a)

(b)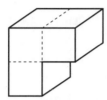

(c)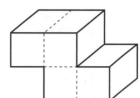

(d)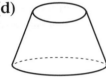

(e)

(f)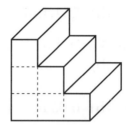

> **Remember:** Plan is the view from the top. Elevations are from the front and the side.

2 Sketch the solid shapes represented by these plans and elevations.

(a)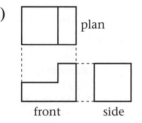

(b)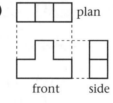

(c)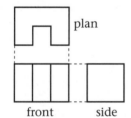

(d)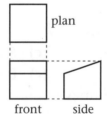

3 Draw two copies of each of these 3-D shapes. On each of them draw a plane of symmetry.

(a)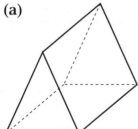

(b)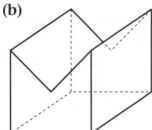

> Draw dashed lines between the plan and elevations to show how they match up.

14 Units of measure

1 Change these lengths to centimetres.

(a) 5 metres　　(b) 11 metres　　(c) 20 mm

(d) 300 mm　　(e) 3.4 m　　(f) 55 mm

(g) 5.7 m　　(h) 254 mm

> **Remember:**
> 10 mm = 1 cm
> 100 cm = 1 m
> 1000 mm = 1 m
> 1000 m = 1 km

2 Change these lengths to millimetres.

(a) 2 cm　　(b) 25 cm　　(c) 3 m

(d) 4.55 m　　(e) 0.5 cm　　(f) 4.56 cm

(g) 0.5 m　　(h) 0.03 m

3 Write these lengths in metres.

(a) 4 km　　(b) 0.5 km　　(c) 4500 cm

(d) 10 000 mm　　(e) 5.43 km　　(f) 55 km

(g) 0.03 km　　(h) 300 cm

4 Change these weights to grams.

(a) 3 kg　　(b) 50 kg　　(c) 450 kg

(d) 2 tonnes　　(e) 0.5 kg　　(f) 0.35 kg

(g) 0.004 kg　　(h) 12.5 kg

> **Remember:**
> 1000 mg = 1 g
> 1000 g = 1 kg
> 1000 kg = 1 tonne

5 Change these weights to kilograms.

(a) 5000 g　　(b) 50 000 g　　(c) 5 tonnes

(d) 30 tonnes　　(e) 450 g　　(f) 0.6 tonnes

(g) 250 g　　(h) 0.003 tonnes

6 Write these weights in tonnes.

(a) 5000 kg　　(b) 600 kg　　(c) 75 kg

(d) 3500 kg　　(e) 39 000 kg　　(f) 75 500 kg

(g) 50 kg　　(h) 55 kg

7 Change these lengths to kilometres.

(a) 4000 m　　(b) 700 m　　(c) 3500 m

(d) 60 000 m　　(e) 500 m　　(f) 235 m

(g) 3450 m　　(h) 50 m

8 Write these volumes in millilitres.

(a) 4 l　　(b) 3.5 l　　(c) 60 l

(d) 0.5 l　　(e) 4.85 l　　(f) 0.06 l

(g) 0.003 l　　(h) 2.05 l

> **Remember:**
> 100 cl = 1 litre
> 1000 ml = 1 litre
> 1000 l = 1 cubic metre

9 Change these volumes into litres.

(a) 5000 ml   (b) 4500 ml   (c) 600 ml

(d) 50 ml   (e) 10 ml   (f) 650 ml

(g) 25 ml   (h) 10 000 ml

10 How many 250 ml glasses can be filled from a 3-litre bottle of cola?

11 How many 2.5 centimetres are there in 150 metres?

Exercise 14.2   Links 14B, D

1 Put these lines in order, shortest first.

2 Write these lengths in order. Put the shortest first.

> Change each length to the same unit.

(a) 5 m, 10 m, 6.5 m, 8 m, 9.5 m, 5.5 m

(b) 400 cm, 5 m, 4.5 m, 475 cm, 550 cm, 5000 mm

(c) 2 cm, 25 mm, 0.1 m, 30 mm, 2.8 cm, 3.4 cm, 18 mm

(d) 300 m, 0.5 km, 250 m, 0.22 km, 500 m, 0.45 km

(e) 250 mm, 20 cm, 0.26 m, 450 mm, 23 cm, 0.5 m

3 Put these weights in order, smallest first.

> Change each weight to the same unit.

(a) 300 g, 0.2 kg, 250 g, 0.4 kg, 500 g

(b) 4000 g, 4.5 kg, 3.5 kg, 3750 g, 4575 g

(c) 675 g, 0.7 kg, 0.6 kg, 650 g, 700 g, 0.75 kg

4 Put these volumes in order, smallest first.

> Change each volume to the same unit.

(a) 200 ml, 300 ml, 0.25 l, 0.275 l, 260 ml

(b) 1 l, 900 ml, 1100 ml, 1.2 l, 1250 ml, 0.75 l

Exercise 14.3 Links 14E, F

The Williams family go on holiday to Devon. The Williams family consist of Mr and Mrs Williams and their three children Lee, Siân and Anna.

Remember:

Metric	Imperial
8 km	5 miles
1 kg	2.2 pounds
25 g	1 ounce
1 l	$1\frac{3}{4}$ pints
4.5 l	1 gallon
1 m	39 inches
30 cm	1 foot
2.5 cm	1 inch

1 Mr Williams works out the distance from their home in Swindon to Devon. He makes it 200 miles.
   Roughly what is this distance in kilometres?

2 Mrs Williams buys two 2-litre bottles of cola for the trip.
   Roughly how many pints is this?

3 Anna eats 4 ounces of sweets on the trip.
   Roughly how many grams is this? Write this amount in kilograms.

4 Siân buys a 500 gram bar of chocolate.
   Roughly how many pounds is that?

5 Mr Williams fills the car up with 20 litres of petrol.
   Roughly how many gallons is that?

6 Mrs Williams puts 500 ml of oil in the engine.
   Roughly how many pints is that?

7 The family stop at a service station 200 kilometres from home. Roughly how many miles is that?

8 Mr Williams puts 2 pints of water in the radiator.
   Roughly how many litres is that?

9 The luggage weighs approximately 150 kg.
   Roughly how many pounds is that?

10 Mrs Williams' luggage weighs 100 pounds.
   Roughly how many kilograms is that?

11 Lee cycles the last 30 miles.
   Roughly how many kilometres is that?

12 Each case is 1 m long by 60 cm wide by 30 cm deep.
   Change these measurements to inches.

Exercise 14.4 Links 14G, H

Remember:
60 seconds = 1 minute
60 minutes = 1 hour
24 hours = 1 day
365 days = 1 year
366 days = 1 leap year
3 months = 1 quarter
12 months = 1 year

1 Change these times into minutes.
   (a) 3 hours   (b) 4 hours   (c) $2\frac{1}{2}$ hours
   (d) $4\frac{1}{2}$ hours   (e) $\frac{1}{2}$ hour   (f) $\frac{3}{4}$ hour
   (g) 3 hours 35 minutes

2 Change these times into hours.

   (**a**) 120 minutes   (**b**) 4 days   (**c**) 30 minutes

   (**d**) 300 minutes   (**e**) $2\frac{1}{2}$ days   (**f**) 200 minutes

   (**g**) 15 minutes   (**h**) 2 days   (**i**) 150 minutes

3 How many seconds are there in

   (**a**) 5 minutes   (**b**) $\frac{1}{2}$ minute   (**c**) $\frac{1}{4}$ minute

   (**d**) 8 minutes   (**e**) 1 hour   (**f**) $2\frac{1}{2}$ hours?

4 Add 20 minutes to each of these times.

   (**a**) 09:00   (**b**) 10:20   (**c**) 11:45   (**d**) 12:50

5 Add 45 minutes to each of these times.

   (**a**) 09:00   (**b**) 10:20   (**c**) 11:45   (**d**) 12:50

6 Change these times into days.

   (**a**) 3 years   (**b**) 2 years and 1 leap year

   (**c**) 120 hours   (**d**) 264 hours

> When you do calculations with time you have to be careful with the carry digit.
> e.g. Add $3\frac{1}{2}$ h to 10:45
>
>   10:45
> \+ 3:30
>   14:15
> 1
>   70 mins makes 1 hour 10 mins.

7 Change these times into hours and minutes.

   (**a**) 3.5 hours   (**b**) $2\frac{1}{2}$ hours

   (**c**) 4.75 hours   (**d**) 3.375 hours

8 Change these times into decimals of an hour.

   (**a**) 3 hours 30 minutes   (**b**) 2 hours 24 minutes

   (**c**) 7 hours 45 minutes   (**d**) 5 hours 6 minutes

9 Add 3 hours 20 minutes to each of these times.

   (**a**) 10:00   (**b**) 12:30   (**c**) 15:40   (**d**) 16:55

10 Subtract 10 minutes from each of these times.

   (**a**) 10:40   (**b**) 12:00   (**c**) 10:10   (**d**) 09:05

11 Subtract 45 minutes from each of these times.

   (**a**) 13:55   (**b**) 16:45   (**c**) 10:30   (**d**) 08:15

12 Subtract 2 hours 30 minutes from each of these times.

   (**a**) 15:50   (**b**) 16:30   (**c**) 14:10   (**d**) 07:05

13 Alice arrives home at 16:30. She watches television for $2\frac{1}{4}$ hours then spends 1 hour 40 minutes on her homework.

   (**a**) At what time does Alice start her homework?

   (**b**) At what time does Alice finish her homework?

Exercise 14.5   Link 14I

Use this part of a calendar to answer questions **1** to **7**.

Days	June	July
Sunday	5 12 19 26	3 10 17 24 31
Monday	6 13 20 27	4 11 18 25
Tuesday	7 14 21 28	5 12 19 26
Wednesday	1 8 15 22 29	6 13 20 27
Thursday	2 9 16 23 30	7 14 21 28
Friday	3 10 17 24	1 8 15 22 29
Saturday	4 11 18 25	2 9 16 23 30

1 What day of the week is 16th June?

2 Which day and date is 3 days after 3rd June?

3 Which day and date is 10 days before 7th July?

4 What is the day and date 14 days after 20th June?

5 What is the day and date a week before 2nd July?

6 What is the date 2 weeks after 17th June?

7 What is the date 10 days after 31st July?

8 Count on 8 days from the following dates.
   (a) 4th Jan
   (b) 3rd Feb
   (c) 7th May
   (d) 25th Aug
   (e) 25th Sept
   (f) 25th Dec

> **Remember:**
> 30 days hath September
> April, June and November.
> All the rest have 31
> except for February alone
> which has just 28 days clear
> and 29 in each leap year.

9 Count back 15 days from the following dates.
   (a) 21st Oct
   (b) 16th July
   (c) 29th Feb
   (d) 8th July
   (e) 7th May
   (f) 6th Jan

Exercise 14.6

Use the two timetables to answer these questions.

Bus Timetable			
Bus station	08:00	09:15	10:30
Stadium	08:15	09:30	10:45
High St.	08:35	09:50	11:05
Hospital	08:45	10:00	11:15
Museum	08:50	10:05	11:20
Bus station	09:10	10:25	11:40

Train Timetable			
Swindon	08:00	09:30	12:45
Kemble	08:15	09:45	13:00
Stroud	08:28	09:58	13:13
Stonehouse	08:40	10:10	13:25
Gloucester	08:55	10:25	13:40
Cheltenham	09:05	10:35	13:50

Remember: On the timetables, times start at the top and go down the page. The time the bus or train leaves is opposite the stopping place.

1 At what time should the 08:00 bus from the bus station be at the High St?

2 At what time should the 08:00 train from Swindon be in Gloucester?

3 At what time was the 10:05 bus from the museum at the stadium?

4 At what time was the 9:58 train from Stroud at Kemble?

5 (a) Buses from the bus station leave every 1 hour 15 minutes. Continue the bus timetable for the next three buses. You may assume that each bus takes the same amount of time between stops as the 08:00 bus.

   (b) Florence arrives at the bus stop in the High St at 11:00. What time is the next bus she could catch to the museum?

6 (a) The next two trains from Swindon leave at 14:15 and 16:05. Continue the train timetable for these next two trains. You may assume that each train takes the same amount of time between stops as the 08:00 train.

   (b) Steven arrives at the train station in Kemble at 09:40. What time is the next train he could catch to Stroud?

7 Work out how long it should take to travel between
   (a) the High St and the museum
   (b) the hospital and the bus station
   (c) the stadium and the museum
   (d) Kemble and Stonehouse
   (e) Stroud and Gloucester.

15 Percentages

1   The large square is divided into 100 small squares.

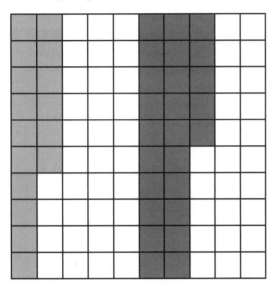

(a)   What percentage of the large square is shaded

       (i)   (ii)   ?

(b)   What percentage is unshaded?

(c)   What fraction is shaded   ?

(d)   Write as decimals the amounts of the large square that are

       (i)   shaded

       (ii)   shaded

       (ii)   unshaded.

2   Change these percentages to fractions in their simplest form.

    (a)   30%   (b)   64%   (c)   57%   (d)   82%

    (e)   75%   (f) $17\frac{1}{2}$%   (g) $33\frac{1}{3}$%   (h)   4%

    (i)   15%   (j) $\frac{1}{4}$%   (k)   28.5%   (l)   3.25%

    (m) $5\frac{1}{2}$%   (n)   82.3%   (o)   6.8%

3 Convert these percentages to decimals.

(a) 40%   (b) 87%   (c) 15%   (d) 4%

(e) 27.8%   (f) 7.8%   (g) $22\frac{1}{2}$%   (h) 0.8%

(i) $\frac{1}{4}$%   (j) $17\frac{1}{2}$%   (k) 58%   (l) 27.5%

(m) 62.4%   (n) 180%   (o) 258%

4 Copy and complete this table of equivalent percentages, decimals and fractions.

Percentage	Decimal	Fraction
60%	0.6	$\frac{3}{5}$
50%		
	0.73	
85%		
$27\frac{1}{2}$%		
	0.25	
		$\frac{2}{5}$
		$\frac{17}{25}$
	0.08	
$2\frac{1}{3}$%		
		$\frac{1}{3}$

5 Change these decimals to percentages.

(a) 0.5   (b) 0.75   (c) 0.25   (d) 0.68

(e) 0.39   (f) 0.41   (g) 0.3   (h) 0.72

(i) 0.15   (j) 0.99   (k) 0.86   (l) 0.31

(m) 0.065   (n) 0.006   (o) 0.001

> Multiply by 100.

6 Write each of these fractions as a percentage.

(a) $\frac{3}{4}$ (b) $\frac{1}{4}$ (c) $\frac{1}{2}$ (d) $\frac{2}{5}$

(e) $\frac{3}{5}$ (f) $\frac{7}{10}$ (g) $\frac{9}{20}$ (h) $\frac{17}{25}$

(i) $\frac{1}{8}$ (j) $\frac{5}{8}$ (k) $\frac{1}{16}$ (l) $\frac{17}{1000}$

(m) $\frac{3}{50}$ (n) $\frac{2}{25}$ (o) $\frac{4}{500}$

> Convert to a decimal and then multiply by 100.

Exercise 15.2

Links 15D–F

1 Work out
 (a) 40% of 275
 (b) 15% of £800
 (c) 12% of 800 m
 (d) 8% of £450
 (e) 10% of £368
 (f) 3% of 6000
 (g) 17.5% of £2800
 (h) 7.8% of 3500 tonnes

2 Increase
 (a) £200 by 10%
 (b) £150 by 20%
 (c) 320 g by 5%
 (d) 250 cm by 4%
 (e) 80 kg by 5%
 (f) 6 tonnes by 25%
 (g) 500 m by $7\frac{1}{2}$%
 (h) £96 by $12\frac{1}{2}$%

3 Decrease
 (a) £300 by 10%
 (b) £250 by 20%
 (c) 450 g by 5%
 (d) 150 cm by 4%
 (e) 180 kg by 5%
 (f) 16 tonnes by 25%
 (g) 400 m by $7\frac{1}{2}$%
 (h) £72 by $12\frac{1}{2}$%

4 The total mark in an examination was 80. The pass mark was set at 60% of the total mark. Work out the pass mark.

5 John invests £450 in a building society. At the end of one year the building society pays John 6.5% interest on the investment. How much interest will John be paid at the end of one year?

> **Remember:** You can't have fractions of a penny, so you may need to round your answer to the nearest penny.

6 The value of a new motor cycle is £3500. When it is 3 years old the motor cycle will have lost 52% of its value when new.
 (a) How much of its value when new will the motor cycle have lost when it is 3 years old?
 (b) Work out the value of the motor cycle when it is 3 years old.

7 An estate agency charges $1\frac{1}{2}$% commission on the selling price of a house. Work out the commission on a house with a selling price of
 (a) £60 000
 (b) £125 000
 (c) £270 000

8 James invests £400 in a building society for one year. The building society pays an interest rate of 4.5% per year.

(a) How much interest, in pounds, does James receive?

His friend, Megan, invests £600 in another building society for a year. At the end of the year she receives £32.40 interest.

(b) Work out the percentage interest rate paid by this second building society.

9 Manuel received a pay rise of 4.5%. Before the pay rise his weekly pay was £240. Work out his weekly pay after the pay rise.

10 Sally wants to buy an MP3 player. The model she wants is advertised in two different shops.

ACE players	MP3s 2 u
£120 less 10%	**£90 plus 15%**

Work out the cost of the MP3 player in each shop.

Exercise 15.3

Link 15G

1 Work out $17\frac{1}{2}$% of these amounts.

(a) £20   (b) £50   (c) £80   (d) £5

(e) £200   (f) £150   (g) £350   (h) £48

(i) £64   (j) £164

2 VAT at 17.5% is added to these amounts. Work out the total amount.

(a) £200   (b) £60   (c) £50   (d) £80

(e) £250   (f) £1500   (g) £800   (h) £2400

(i) £480   (j) £4000

3 Value added tax is charged at the rate of $17\frac{1}{2}$%.
Work out the total cost of these bills.

(a) A mobile phone bill of £32 before VAT is added.

(b) A garage bill of £120 before VAT is added.

(c) A washing machine repair bill of £60 before VAT is added.

(d) A computer repair costing £56 before VAT is added.

(e) A plumber's bill of £300 before VAT is added.

4 A van is advertised for sale. The advertisement reads

> *For Sale* **£3200 plus VAT**

VAT is charged at 17.5%.

(a) Work out the VAT that will be charged on the van.

(b) Work out the total selling price of the van when VAT is added.

Exercise 15.4   **Link 15H**

1 The table shows the index number for the value of a house in Swindon.

> **Remember:** An index always starts at 100.

Year	1975	1985	1995	2005	2015
Index	100	200	150	225	

(a) What was the percentage change from 1975 to 1985?

(b) In what period was the greatest change in house value?

(c) Why is it not possible to estimate the index value in 2015?

2 This table shows the index number for the price of petrol in one month during a petrol crisis.

Week	1	2	3	4
Index	100	105	120	110

(a) What was the percentage change from Week 1 to Week 3?

(b) In what period was the greatest change in petrol price?

(c) What is the overall trend in the price of petrol?

3 This table shows the index number for the number of people who took part in free-fall parachute jumps in the last 5 years.

Year	1	2	3	4	5
Index	100	105	115	130	150

(a) What was the percentage change from Year 3 to Year 5?

(b) In what period was the greatest change in the number of people taking part in a free-fall parachute jump?

(c) What is the overall trend in the number of people taking part in a free-fall parachute jump?

4 Here is a table that shows the index number for the number of smokers in Bristol over the last five years.

Year	1	2	3	4	5
Index	100	95	92	90	89

(a) What was the percentage change from Year 1 to Year 3?

(b) In what period was the greatest change in the number of smokers?

(c) What is the overall trend in the number of smokers in Bristol?

Exercise 15.5   Links 15I, J

1 Asif paid a deposit of £384 on a new motor cycle with a cost price of £3200. What percentage of the cost price was the deposit?

2 In a sale, the price of a coat was reduced from £110 to £82.50. What was the percentage reduction in the price of the coat?

3

> **Equate Computer System**
>
> Normal Price   £1600
> Sale Price   £1360

Work out the percentage reduction on the Equate Computer in the sale.

4 Write
(a) 16 as a percentage of 40
(b) 12 as a percentage of 30
(c) £300 as a percentage of £500
(d) £250 as a percentage of £400.

5 (a) What percentage of £200 is £24?
(b) What percentage of 1500 m is 300 m?
(c) Write £600 as a percentage of £750.
(d) Write 40 cm as a percentage of 85 cm.

6 The normal price of a coat is £60.
The sale price of this coat is £36.
Write the sale price as a percentage of the normal price.

7   Glassport's employ 1650 people. 395 of these people are left-handed. What percentage of the people employed by Glassport's are left-handed? Give your answer to the nearest whole number.

8   Frances sees three different advertisements for jeans.

Work out the cost of the jeans in each advertisement.

9   One mile = 1600 metres (approximately).
Write one kilometre as a percentage of a mile.

10   In a sale, the price of a pair of shoes was reduced from £60 to £51. Work out the reduction as a percentage of the original price.

11   Jim and Lesley bought a flat for £65 000. They sold it a year later for £76 000. Work out the profit they made as a percentage of the amount they paid for the flat.

12   In 1961 the average price of a litre of petrol was 2.5p.
In 2006 the average price of a litre of petrol was 97.5p.
Work out the percentage increase in the price of a litre of petrol over the 45 years.

13   Re-arrange these numbers in order of size, starting with the smallest.
   (a) 53%, 0.52, $\frac{9}{15}$ **(b)** 83%, $\frac{4}{5}$, 0.82
   (c) 0.006, $\frac{6}{100}$, 6.1%   **(d)** $\frac{4}{5}$, 80%, 0.8

16 Coordinates and graphs

1

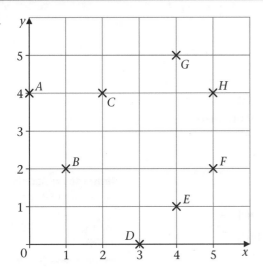

> **Remember:** The number of units across is always first and the number of units up is second.

(a) Write down the letters on the grid with these coordinates.

    (i)   (0, 4)   (ii) (4, 5)   (iii) (1, 2)   (iv) (5, 2)

(b) Write down the coordinates of these letters on the grid.

    (i)   *D*   (ii) *H*   (iii) *C*   (iv) *E*

2 Write down the coordinates of all the points marked that make up this shape.

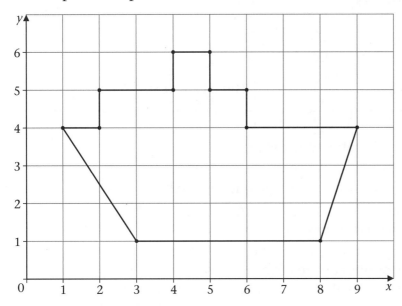

3 Draw a coordinate grid from 0 to 10 in both directions. On your grid plot the points and draw the following shapes by joining the points in order.

(a) (1, 1), (1, 6), (2, 7), (7, 7), (8, 6), (8, 1), (1, 1)

(b) (4, 1), (4, 3), (5, 3), (5, 1)

(c) (4, 7), (4, 9), (6, 9), (6, 7)

(d) (2, 4), (2, 6), (4, 6), (4, 4), (2, 4)

(e) (5, 4), (5, 6), (7, 6), (7, 4), (5, 4)

4 Draw a coordinate grid from 0 to 10 in both directions. On your grid draw a design of your own. Label each point and write down its coordinates.

5 Work out the coordinates of the mid-point of the line segment between each pair of points.

(a) A(0, 0) B(6, 8) **(b)** C(2, 5) D(4, 8)

(c) E(3, 7) F(4, 11) **(d)** G(1, 2) H(9, 7)

> **Remember:** Add the x-coordinates and divide by 2.
> Add the y-coordinates and divide by 2.

Exercise 16.2   Links 16C, D

1 The table shows the cost of Brussels sprouts per kg.

Weight in kg	1	2	3	4	5
Cost in pence	80	160	240	320	400

(a) Draw a graph for this table.

(b) Work out the cost of
  (i) $2\frac{1}{2}$ kg   **(ii)** $4\frac{1}{2}$ kg

(c) Extend the graph to work out the cost of
  (i) 6 kg   **(ii)** 8 kg   **(iii)** 10 kg

2 The table shows the distance travelled by a four-wheel drive vehicle for every gallon of diesel it uses.

Distance travelled (miles)	25	50	75	100	125	150
Diesel used (gallons)	1	2	3			

(a) Copy and complete the table to show how much diesel was used.

(b) Draw a graph from the information given in your table.

(c) Work out how much diesel was used in travelling
  (i) 40 miles   **(ii)** 110 miles

(d) Work out how many miles were travelled when the following amounts of diesel were used.
  (i) 8 gallons   **(ii)** 10 gallons   **(iii)** 20 gallons

3 (a) Draw a conversion graph from litres to pints.
Use these facts:

    0 litres = 0 pints and 50 litres = 90 pints

On your graph use these scales for litres and pints:

    1 cm = 5 litres and 1 cm = 5 pints

Plot the points (0, 0) and (50, 90) and join them with a straight line.

(b) Copy and complete this table using your conversion graph to help you.

Litres	0			38		25	44	28	15			50
Pints	0	10	20		36					50	72	90

4 Copy this table which converts ounces to grams. Use the information to draw a conversion graph from ounces to grams. Use your graph to help fill in the missing values.

Ounces (oz)	1	2			16	10	24	19
Grams (g)	28		113	340			680	

5 Use your graph in question **4** to find the number of grams for 8 ounces.

6 (a) Draw a graph using the information below to show the speed of a car and distance travelled in a drag race.

Speed in km/hr	0	25	60	150	190	250
Distance in metres	0	80	250	600	800	1000

(b) Use your graph to work out the speed when the distance travelled is 500 m.

(c) Use your graph to work out the distance travelled when the speed is 100 km/hr.

Exercise 16.3   Links 16E, F

1 June went shopping by van. She drove 15 miles to the shops in 20 minutes. She stayed at the shops for 45 minutes and then started to drive home. After 10 minutes, when she had driven 5 miles, she stopped for petrol for 5 minutes. It then took her a further 15 minutes to get home.

Draw a distance–time graph for June's journey.

> **Remember:** Distance is along the vertical axis. Time is along the horizontal axis.

2 Mustaq walks to the café, meets some friends, then walks home again.

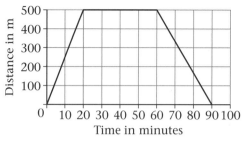

(a) How many minutes did it take Mustaq to walk to the café?

(b) How far away was the café?

(c) How many minutes did Mustaq spend at the café?

(d) How long did it take Mustaq to walk home?

(e) Work out the speed, in metres per minute, for Mustaq to walk to the café.
Also give your answer in kilometres per hour.

(f) Work out the speed, in metres per minute, for Mustaq to walk back from the café. Give your answer in kilometres per hour as well.

> **Remember:**
> Speed = distance travelled ÷ time taken

> 1 km = 1000 m
> 1 hour = 60 minutes
> 1 minute = 60 seconds

3 Marion travels to school by coach. She walks the first 500 metres in 10 minutes, waits at the bus stop for 6 minutes, then travels the remaining 2500 metres by coach. She arrives at the school coach stop 21 minutes after she set off from home.

(a) Draw a distance-time graph of her journey.

(b) Work out the speed of the coach, first in metres per minute, then in kilometres per hour.

Exercise 16.4   Link 16G

1 Write down the coordinates of the points labelled *A* to *L* on the coordinate grid.

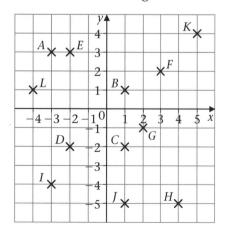

2 Draw a coordinate grid with the horizontal axis (x-axis) marked from -10 to $+10$ and the vertical axis (y-axis) marked from -5 to $+5$.

Plot the following points and join them in the order of this list.

$(-10, 0)$, $(-10, 2)$, $(-5, 4)$, $(-4, 5)$, $(-2, 1)$, $(2, 3)$,
$(4, 2)$, $(9, 3)$, $(10, 2)$, $(10, -2)$, $(9, -3)$, $(4, -2)$, $(2, -4)$,
$(-2, -1)$, $(-4, -5)$, $(-5, -4)$, $(-10, -2)$, $(-10, 0)$

Exercise 16.5   Links 16H–J

1 (a) Copy and complete the tables of values for the equations below.

(i) $y = 4x + 1$

x	-2	-1	0	1	2
y					

(ii) $y = 3 - 2x$

x	-2	-1	0	1	2
y					

(b) Draw the two graphs on the same grid and write down the coordinates of the point where they cross.

2 Write down the equations of the lines marked **(a)** to **(e)** in this diagram.

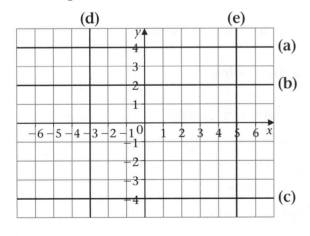

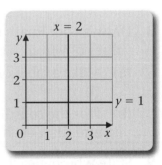

3 Draw a coordinate grid with x- and y-axes labelled from -5 to $+5$. On the grid draw and label the graphs of

(a) $x = 1$ **(b)** $x = 3$ **(c)** $x = -5$ **(d)** $x = -2$

(e) $y = 3$ **(f)** $y = -1$ **(g)** $y = 1$ **(h)** $y = -3$

4 Draw a coordinate grid with axes labelled from -5 to $+5$.
On the grid draw and label the graphs of

(a) $y = -3$

(b) $x = +2$

(c) Write down the coordinates of the point where the two lines cross.

5 On a coordinate grid with the x-axis labelled from -5 to $+5$ and the y-axis labelled -5 to $+5$ draw the following graphs.

(a) $y = x$ (b) $y = x + 3$ (c) $y = x - 3$

(d) $y = x + 5$ (e) $y = x - 5$

6 On a coordinate grid with the x-axis labelled -4 to $+4$ and the y-axis labelled -12 to $+12$ draw the following graphs.

(a) $y = 3x$ (b) $y = 3x + 8$

(c) $y = -3x + 6$ (d) $y = -2x - 4$

7 (a) Copy and complete the tables of values for the equations below.

$y = 2x - 3$

x	-2	-1	0	1	2	3	4
y							

$y = -x + 2$

x	-2	-1	0	1	2	3	4
y							

(b) Draw the graphs on graph paper.

(c) Write down the coordinates of the point where they cross.

8 Draw a coordinate grid with values of x from -4 to $+4$ and values of y from -6 to $+12$.

(a) Make a table of values for the equations below.
(i) $y = x^2 - 5$ (ii) $y = 2x + 3$

(b) Plot the coordinates for each of the tables of values and draw each graph.

9 State whether each shape is 1-D, 2-D or 3-D.

(a) octagon   (b) rectangle   (c) cuboid

10 Write down all the 3-dimensional coordinates from this list.

(123), (3, 5, 7), (9, 3), (7), (25, 6), (2, 3, 4), (9, 2, 3), (3)

17 Ratio and proportion

1 Here are some tile patterns. For each one write the ratio of white tiles to shaded tiles.

(a)

(b)

(c)

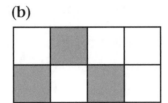

2 Draw a copy of this diagram.

Shade your diagram so that 3 parts of it are shaded and the rest is unshaded.

3 A recipe for 4 cakes needs 50 g of butter and 120 g of flour. Work out

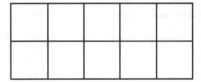

8 cakes need twice as much ingredients as 4 cakes.

(a) how much butter is needed for 8 cakes

(b) how much flour is needed for 8 cakes

(c) how much flour is needed for 12 cakes

(d) how much butter is needed for 6 cakes

(e) how much butter is needed for 18 cakes

(f) how much flour is needed for 10 cakes

(g) how much butter is needed for 10 cakes.

4 A builder uses 3 buckets of sand and 1 bucket of cement to make concrete.

(a) How many buckets of sand will he need when he uses 12 buckets of cement?

(b) How many buckets of cement will he need when he uses 15 buckets of sand?

5 A drink is made by mixing orange juice and water in the ratio 1 : 4.
How much orange juice would need to be used to make a drink which contained 60 ml of water?

6 Write these ratios in their lowest terms.

(a) 3 : 6   (b) 2 : 10   (c) 4 : 32   (d) 4 : 10

(e) 12 : 4   (f) 18 : 9   (g) 30 : 12   (h) 35 : 15

> **Remember:** Ratios cancel down like fractions.
> $100 : 60 = 50 : 30 = 5 : 3$
> They have common factors.

7 Write as ratios in their lowest terms.

(a) 20p : 100p   (b) 30p : £1

(c) £5 : 50p   (d) 2 metres : 10 cm

(e) 5 cm : 1 metre   (f) 12 minutes : 1 hour

8 Write these ratios in their simplest form.

(a) 30 : 20   (b) 18 : 6   (c) 24 : 15

(d) 10 : 4   (e) 36 : 24   (f) 8 : 20

(g) 12 : 14   (h) 9 : 48   (i) 143 : 1001

(j) 5 : 10 : 15   (k) 48 : 24 : 6   (l) 14 : 56 : 35

9 Write these ratios in their simplest form.

(a) 2 cm : 1 m   (b) 450 mg : 1 g

(c) 25 ml : 2 l   (d) 2 kg : 50 g

(e) 48p : £2   (f) 40 s : 1 minute

(g) 55 cm : 2 m   (h) 12 mm : 1 m

(i) £2.40 : £3.60   (j) 1 mm : 1 cm : 1 m

(k) 82p : £5   (l) 20 minute : 2 hours : $\frac{1}{4}$ hour

> **Remember:** Measurements must be in the same units.

Exercise 17.2   Links 17C–D

1 Calculate the missing numbers in these ratios.

(a) 4 : 5 = 12 : ?   (b) 5 : 8 = 30 : ?   (c) 2 : 7 = 6 : ?

(d) ? : 3 = 7 : 6   (e) ? : 7 = 16 : 28   (f) ? : 1 = 16 : 10

2 The diagram represents a lamp-post and the shadow it casts at noon.

On the same day at noon

(a) how long is the shadow cast by a tree which is 20 metres tall

(b) how long is the shadow cast by a man who is 2 metres tall

(c) how tall is a building which casts a shadow of length 60 metres?

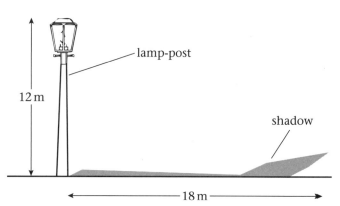

3 Write these ratios in their simplest form.

(a) $3 : \frac{1}{4}$ (b) $4 : \frac{1}{2}$ (c) $\frac{3}{4} : \frac{1}{2}$

(d) $1\frac{1}{2} : 2$ (e) $\frac{1}{3} : \frac{1}{2}$ (f) $\frac{3}{4} : 2\frac{1}{4}$

(g) $\frac{3}{4} : \frac{5}{8} : \frac{1}{2}$ (h) $\frac{3}{5} : \frac{7}{8}$ (i) $\frac{3}{4} : 3\frac{1}{2} : 2$

4 Find x for each of these pairs of equivalent ratios.

(a) $x : 4$ $16 : 24$ (b) $4 : 13$ $20 : x$ (c) $x : 6$ $14 : 42$

(d) $18 : 6$ $x : 1$ (e) $9 : 45$ $1 : x$ (f) $9 : x$ $45 : 80$

(g) $\frac{1}{2} : x$ $10 : 20$ (h) $3 : 8$ $x : 5$ (i) $5 : x$ $35 : 63$

5 Write each of these ratios in the form $1 : n$.

(a) $2 : 7$ (b) $3 : 15$

(c) $6 : 15$ (d) $5 : 13$

(e) $20\,\text{cm} : 2\,\text{m}$ (f) $40\,\text{g} : 1\,\text{kg}$

(g) $45\text{p} : £2$ (h) $56\text{p} : £3$

(i) $30\,\text{minutes} : 2\,\text{hours}$

6 Write each of these ratios in the form $n : 1$.

(a) $20 : 4$ (b) $42 : 5$

(c) $6 : 12$ (d) $6 : 15$

(e) $8 : 6$ (f) $4\,\text{hours} : \frac{1}{2}\,\text{hour}$

(g) $2\,l : 10\,\text{m}l$ (h) $4\,\text{m} : 2\,\text{cm}$

(i) $50\text{p} : £2$

7 The ratios $3 : 5 : x$ and $24 : 40 : 56$ are equivalent. Find x.

Exercise 17.3   **Links 17E**

1 Divide the quantities in the ratios given.

> **Remember:** $5 : 4$ means 5 parts to 4 parts. So there are 9 parts in total.

(a) £16 in the ratio $3 : 5$ (b) £24.80 in the ratio $2 : 3$

(c) £130 in the ratio $10 : 3$ (d) £726 in the ratio $1 : 2 : 3$

(e) 3 m in the ratio $3 : 7$ (f) 45 cm in the ratio $4 : 5$

(g) 96 kg in the ratio $5 : 3$ (h) £360 in the ratio $7 : 11$

(i) 61.56 m in the ratio $5 : 4$ (j) £721 in the ratio $1 : 4 : 2$

2 The ratio of girls to boys in a class is $7 : 6$.
There are 26 students in the class.
Find out how many are **(a)** girls **(b)** boys.

3 Elizabeth and Sasha share £40 in the ratio $3 : 5$.
How much should

(a) Elizabeth receive   (b) Sasha receive?

4 There are 1300 students at Lucea College.
The ratio of male to female students at the
college is 6 : 7.
How many female students are there at
Lucea College?

5 Mr Mohammed won £3600 on the Lottery.
He shared the money between himself, his wife and his
son in the ratio 1 : 3 : 2.
Work out each person's share of the money.

6 Mortar is made by mixing weights of sand and cement in
the ratio 7 : 2.

   (a) How much sand is needed to make 1800 kg of mortar?

   (b) How much cement is needed to make 4500 kg of
   mortar?

7 The ratio of girls to boys in a drama group is 7 : 6.
There are 24 boys in the drama group.
How many girls are there in the drama group?

> When the ratio is $a : b$, is
> the value given in the
> question for a or b?

8 The ratio of the length to the width of a rectangle is 5 : 4.
The width of the rectangle is 12 cm.
Work out the length of the rectangle.

9 A drink contains orange juice and water in the ratio 1 : 4.
There are 340 ml of water in the drink.
Work out the amount of orange juice in the drink.

10 A concrete mix is made by adding sand and cement in the
ratio 5 : 1.
Six buckets of cement are put into a mixer.
Work out the number of buckets of sand that should be
put into the mixer.

Exercise 17.4   Links 17F, G

1 Zorba bought 12 litres of petrol for his motorbike.
The total cost of the petrol was £9.06.

   (a) What was the cost of 1 litre of petrol?

   Eric bought 16 litres of the same petrol for his car.

   (b) How much did Eric pay for this petrol?

> Parts **(a)** and **(b)** in
> question **1** show the
> method to use in these
> questions.

2 Barbara is paid £18 for four hours' work.

    (a) How much is she paid for each hour?

    (b) How much should she be paid for
        (i) six hours' work
        (ii) fifteen hours' work?

3 The Head of English at Lucea High School buys 36 copies of *Great Expectations* for £234.
The Head of English at Russell High School buys 50 copies of *Great Expectations*.
How much should the Head of English at Russell High School pay for these books?

4 Six standard size jars of coffee cost £13.20. Work out the cost of eight standard size jars of the same coffee.

5 Five identical tubes of toothpaste have a total capacity of 600 m*l*. Work out the total capacity of nine of these tubes of toothpaste.

6 When it is full, a trough provides enough drinking water to last 12 horses for 4 days. How many days will the same trough, when full, last for

    (a) 9 horses

    (b) 18 horses

    (c) 25 horses?

7 Six identical machines can complete a job in 5 hours. How long will it take

    (a) one machine to complete the job

    (b) twelve machines to complete the job

    (c) four machines to complete the job?

8 Using three identical copiers, a printer can make a copy of a manuscript in 20 minutes.
How long will it take the printer to make a copy of the same manuscript if she has

    (a) only two identical copiers

    (b) five copiers?

9 A quantity of food will last six men for nine days. How long will the same quantity of food last

    (a) 4 men   **(b)** 18 men   **(c)** 5 men?

10 Joan works for 4 hours and is paid £18.
Sam earns the same amount for each hour's work as Joan.
How much should Sam be paid for 10 hours' work?

11 Six men can build a boat in 24 days.
Working at the same rate, how long should it take

(a) 12 men to build the same sort of boat

(b) 9 men to build the same sort of boat?

12 Ten litres of central heating oil cost £1.62.
Work out the cost of

(a) 1 litre of this oil

(b) 100 litres of this oil

(c) 750 litres of this oil.

13 A packet of 20 Christmas cards costs £3.20.

(a) Work out the cost of each Christmas card.

(b) Work out the cost of 12 of these packets of Christmas cards.

14 Asif works as a sales representative. He receives 35p for each mile he travels on business as travel expenses. Work out how much he will receive as travel expenses if he travels

(a) 10 miles   (b) 30 miles   (c) 250 miles

Last year Asif travelled 8400 miles on business.

(d) How much did he receive as travel expenses?

15 The cost of hiring a boat for a week is £1560.
A group of friends share costs equally between them.
How much will it cost each person if the boat is

Work out how much it costs for 1 person.

(a) hired for a week and there are
   (i)   4 people in the group
   (ii) 10 people in the group?

(b) hired for 2 weeks and there are
   (i)   7 people in the group
   (ii) 8 people in the group?

Exercise 17.5   Link 17H

1 A map has a scale of 1 : 20 000. What is the distance on the ground if the distance on the map is

Remember: 1 : 20 000 means 1 cm represents 20 000 cm

(a) 2 cm   (b) 3.5 cm
(c) 14.2 cm   (d) 25.8 cm?

2 An architect draws a plan of a house. The plan is drawn to a scale of 1 : 25. Work out

    (a) the height of the house on the plan given that the true height of the house is 12.5 metres

    (b) the length of the house given that the length of the house on the plan is 64 cm

    (c) the width of the house on the plan given that the true width of the house is 14.6 metres.

3 The real-life distance between Lucea and Presswell is 24 kilometres. The distance between Lucea and Presswell on a map is 6 cm.

    (a) On the map, how many kilometres are represented by 1 cm?

    (b) Work out the real-life distance between two towns which are 8 cm apart on the map.

    (c) Work out the scale of the map as a ratio in its lowest terms.

4 A model boat has a scale of 1 cm to represent 1.2 metres.

    (a) Write this scale as a ratio.

    (b) What is the length of the model if the length of the real boat is 24 metres?

    (c) Work out the width of the real boat if the width of the model is 14 cm.

5 A model castle has a scale of 1 : 40.

    (a) Work out the height of the dining room in the castle given that the height of the dining room in the model is 15.4 cm.

    (b) Work out the length of the drawbridge on the model castle given that the length of the real drawbridge is 8 metres.

    (c) Work out the perimeter of the moat around the castle given that the perimeter of the moat on the model is 6.48 metres.

6 Caesar was driving along Ermine Street and measured the distance he travelled in a straight line as 12 km. What distance would that be on a map with a scale of 1 : 50 000?

18 Symmetry

1 For each design write down how many lines of symmetry there are.

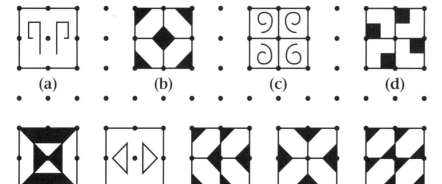

(a)   (b)   (c)   (d)

Remember: If you could fold each design in half and one half fits **exactly** on top of the other, the shape has a line of symmetry.

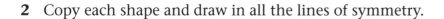

(e)   (f)   (g)   (h)   (i)

2 Copy each shape and draw in all the lines of symmetry.

Using tracing paper is a good way to find lines of symmetry.

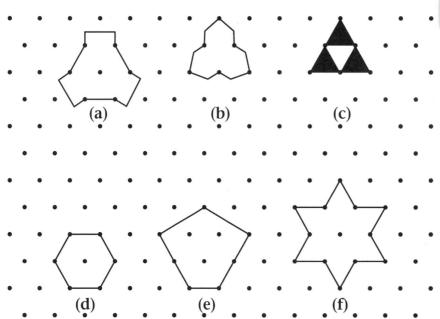

(a)   (b)   (c)

(d)   (e)   (f)

3 Using the dotted lines as lines of symmetry, copy and complete each shape.

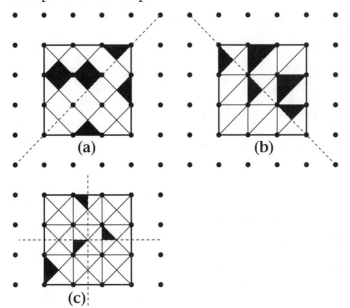

(a)   (b)

(c)

1 Which of these shapes have rotational symmetry?

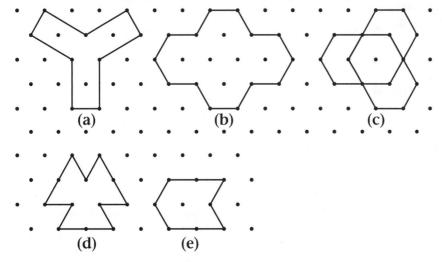

(a)   (b)   (c)

(d)   (e)

> To help do these questions turn the book or use tracing paper.

> **Remember:** A 2-D shape with rotational symmetry repeats the appearance of its starting position two or more times during a full turn.

2 Write down the order of rotational symmetry for each shape in question **1**.

> **Remember:** The order of rotational symmetry is the number of times the original position is repeated in a full turn.

3 You can form shapes with rotational symmetry of order 3 by modifying equilateral triangles.

Here are two examples of how you can do this.

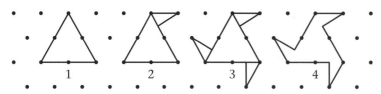

1 Start with an equilateral triangle.
2 Add something extra to one side.
3 Add the same to the other two sides.
4 Redraw with only the outline.

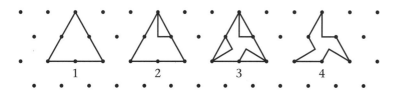

Make some shapes of your own with rotational symmetry of order 3.

4 Use the same idea with a square to make shapes with rotational symmetry of order 4.

5 For regular polygons with
(i) 10 sides   **(ii)** 9 sides
state

(a) the number of lines of symmetry

(b) the order of rotational symmetry.

19 Simple perimeter, area and volume

Exercise 19.1
Links 19A–C

1 Work out the perimeters of these shapes.

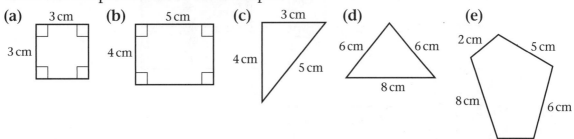

2 Work out the perimeters of these shapes.

(a) a square with side 6 cm

(b) a rectangle with sides 6 cm and 5 cm

(c) an equilateral triangle with side 10 cm

(d) a regular pentagon with side 4 cm

(e) an isosceles triangle with two sides of 4 cm and one of 6 cm

> **Remember:** The perimeter is the distance around the outside of a shape.

3 Measure these shapes and then find their perimeters.

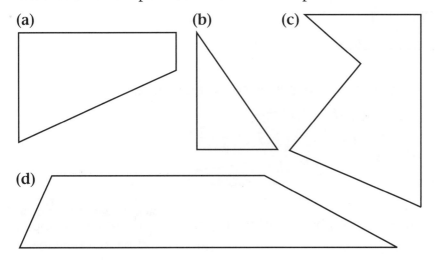

4 Find the perimeters of these shapes.

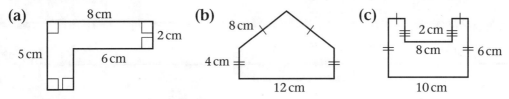

Exercise 19.2

Links 19D–F

1 Find the areas of these shapes by counting squares.

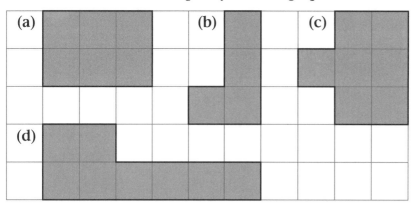

> **Remember:** Area is the number of squares inside a shape.

2 Estimate the areas of these shapes by counting squares.

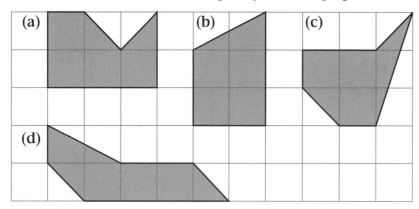

3 Estimate the areas of these curved shapes by counting squares.

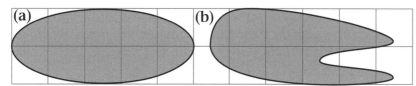

> **Remember:** Volume is the space inside a solid shape.

4 Find the volumes of these shapes by counting cubes.

(a) 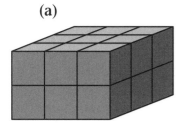 **(b)** 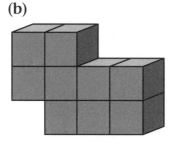 **(c)**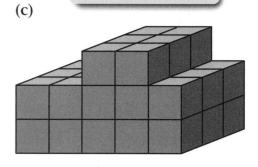

Exercise 19.3

1 Find the areas of these shapes.

(a)
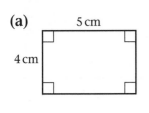

(b)
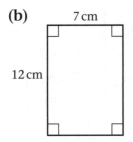

(c)
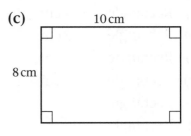

(d)
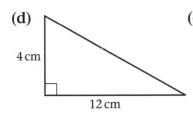

(e)
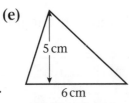

(f)
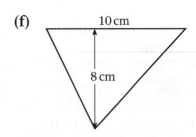

> **Remember:**
>
>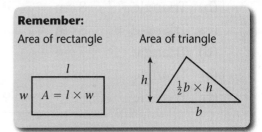
>
> Area of rectangle   Area of triangle
>
> $A = l \times w$ $\frac{1}{2}b \times h$

2 Find the areas of these composite shapes.

(a)
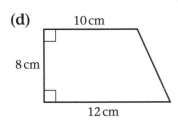

(b)
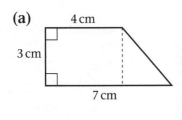

(c)
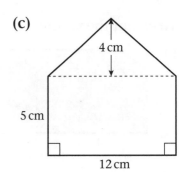

(d)
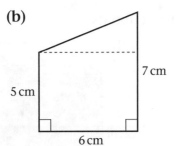

(e)
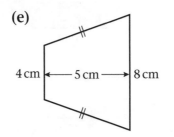

(f)
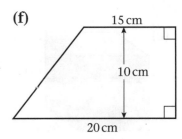

3 Copy this table into your book and complete the missing measurements.

Shape	Length	Width	Area
(a) Rectangle	4 cm	5 cm	
(b) Rectangle	5 cm	8 cm	
(c) Rectangle	8 cm		32 cm²
(d) Rectangle	7 cm		28 cm²
(e) Rectangle		2 cm	16 cm²
(f) Rectangle		9 cm	108 cm²

4 Copy this table into your book and complete the missing measurements.

Shape	Base	Vertical height	Area
(a) Triangle	6 cm	5 cm	
(b) Triangle	8 cm	6 cm	
(c) Triangle	8 cm		16 cm²
(d) Triangle	7 cm		21 cm²
(e) Triangle		2 cm	12 cm²
(f) Triangle		9 cm	36 cm²

Exercise 19.4   **Links 19I–L**

1 Work out the volumes of these cuboids.

(a)
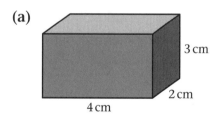
3 cm
2 cm
4 cm

(b)
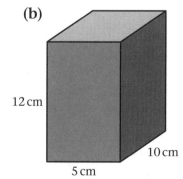
12 cm
10 cm
5 cm

Remember: The volume of a cuboid is
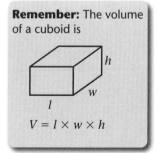
h
w
l
$V = l \times w \times h$

(c)
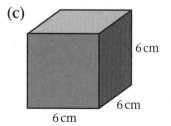
6 cm
6 cm
6 cm

2 Copy the table into your book and fill in the missing measurements.

	Shape	Length	Width	Height	Volume
(a)	Cuboid	6 cm	4 cm	2 cm	
(b)	Cuboid	5 cm	4 cm	3 cm	
(c)	Cube	3 cm	3 cm	3 cm	
(d)	Cuboid	6 cm		3 cm	36 cm^3
(e)	Cuboid	8 cm		2 cm	48 cm^3
(f)	Cuboid		4 cm	2 cm	24 cm^3
(g)	Cuboid		5 cm	3 cm	75 cm^3
(h)	Cuboid	4 cm	2 cm		32 cm^3
(i)	Cuboid	10 cm	5 cm		100 cm^3
(j)	Cube	4 cm			64 cm^3
(k)	Cube				1000 cm^3

3 How many cubes of side 2 cm will fit into a cuboid measuring 10 cm by 8 cm by 6 cm?

4 Work out the surface area of a cube with sides of length 5 cm.

> A cube has 6 faces. Each face is a square.

5 Work out the surface area of these shapes.

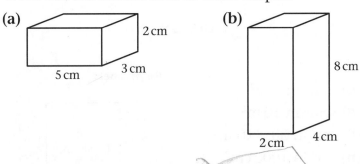

(a) 2 cm, 3 cm, 5 cm

(b) 8 cm, 2 cm, 4 cm

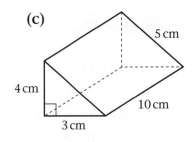

(c) 5 cm, 4 cm, 10 cm, 3 cm

6 How many boxes of cereal measuring 30 cm by 20 cm by 10 cm will fit into a carton measuring 90 cm by 60 cm by 30 cm?

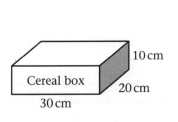

Cereal box
10 cm
20 cm
30 cm

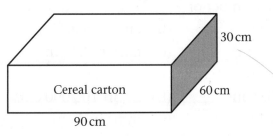

Cereal carton
30 cm
60 cm
90 cm

7 How many cubes of sugar with edge length 1 cm will fit into a box measuring 15 cm by 10 cm by 4 cm?

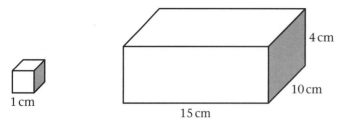

8 A box of matches measures 5 cm by 3 cm by 1 cm. How many boxes of matches will fit into a box measuring 30 cm by 12 cm by 10 cm?

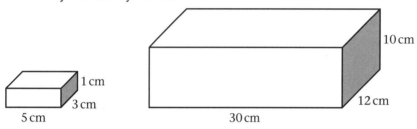

9 Bill and Ben decide to put up a greenhouse on a rectangular concrete base. The base for the greenhouse is in the shape of a cuboid which is 2.6 m long, 1.9 m wide and 15 cm deep. Concrete costs £50 per cubic metre. Work out the cost of the concrete base for the greenhouse.

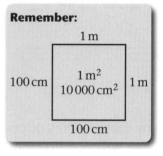

Exercise 19.5   **Link 19M**

1 Work out the number of
   (a) cm^2 in 3 m^2 **(b)** cm^2 in 4 m^2
   (c) cm^2 in 3.5 m^2 **(d)** cm^2 in 10 m^2
   (e) m^2 in 100 000 cm^2 **(f)** m^2 in 30 000 cm^2
   (g) cm^2 in 0.05 m^2 **(h)** m^2 in 1000 cm^2

Remember:

2 Work out the number of
   (a) cm^3 in 3 m^3 **(b)** cm^3 in 5 m^3
   (c) cm^3 in 2.5 m^3 **(d)** cm^3 in 3.47 m^3
   (e) m^3 in 6 000 000 cm^3 **(f)** m^3 in 50 000 000 cm^3
   (g) cm^3 in 0.05 m^3 **(h)** m^3 in 100 000 cm^3
   (i) mm^3 in 1 m^3 **(j)** mm^3 in 5 cm^3

Remember:
$1 m^3 = 100 cm \times 100 cm \times 100 cm$
$= 1 000 000 cm^3$

3 The volume of a garden shed is 6 m^3.
How many cm^3 is this?

4 The area of one page in a book is 40 cm^2.
There are 100 pages in the book.
Write the total area of paper in m^2.

Exercise 19.6

Link 19N

1 Catherine walked for 2 hours at 4 km per hour.
How far did she walk?

2 Keith drove for 4 hours at an average speed of 60 miles per hour.
How far did he drive?

3 Peter cycled at 12 miles per hour for 3 hours.
How far did he cycle?

4 Steve drove 100 kilometres in 2 hours.
At what average speed did he drive?

5 Barry walked 15 miles in 5 hours.
At what average speed did he walk?

6 Karen travelled 48 kilometres in 3 hours.
At what average speed did she travel?

7 Dave drove 120 miles at 60 miles per hour.
How long did it take him?

8 Marlene walked 10 miles at 4 miles per hour.
How long did it take her?

Remember:

You can use this triangle to help you remember the formula connecting Distance, Speed and Time

20 Presenting and analysing data 1

Exercise 20.1 Links 20A–D

1 Find the mode, median, mean and range of these sets of data.
 (a) 4, 6, 9, 8, 7, 9, 10, 4, 7, 6, 9, 9
 (b) 6, 9, 9, 10, 10, 7, 8, 7, 6
 (c) 5, 5, 5, 5, 5, 5, 5, 5, 5, 5
 (d) 5.5, 5.6, 5.7, 5.3, 5.4, 5.5, 5.2
 (e) 12, 15, 11, 17, 14, 16, 18, 13, 12, 15
 (f) 45, 47, 45, 44, 42, 47, 45, 43, 44
 (g) 65, 67, 65, 64, 62, 67, 65, 63, 64
 (h) 67, 45, 32, 54, 55, 76, 67, 74, 67, 64, 55, 67

2 Myra was preparing a mail-shot and recorded the weights of the packages she was sending. The weights of the first 6 packages were

  51 g 55 g 54 g 55 g 56 g 50 g

 (a) Work out the median. **(b)** Work out the mean.

 Myra then weighed another 4 packages. Their weights were

  55 g 59 g 60 g 55 g

 (c) Work out the mean weight of these 4 packages.

 (d) Work out the mean weight of all 10 packages

3 Selena is looking for a job. She puts her CV on the internet. The bar chart shows information about the numbers of enquiries she gets.

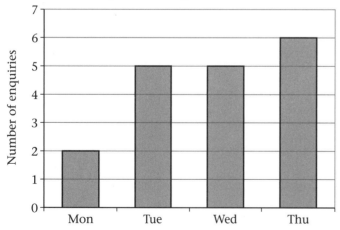

 (a) Work out the mode. **(b)** Work out the mean.
 Selena got 7 enquiries on Friday.
 (c) Work out the mode and mean for all 5 days.

Remember:	
Mode	The value that occurs most often. (It is possible to have more than one mode.)
Median	The middle value when the data is arranged in order of size.
Mean	The sum of the values divided by the number of values.
Range	The difference between the highest and lowest values.

4 Justine measured the heights of the 10 girls in her kick boxing class. Here are the results.

    1.80 m,   1.70 m,   1.65 m,   1.72 m,   1.82 m
    1.65 m,   1.75 m,   1.65 m,   1.74 m,   1.73 m

  (a) Work out the mean height.

  (b) Work out the range of the heights.

Another girl joined the class. When her height was included the mean height increased by 1 cm.

  (c) How tall was the new girl?

5 Abbas worked out the mean weight of 10 blocks of cheese. It worked out to be 250 grams. When he added another block of cheese the mean decreased by 1 gram. How heavy was the eleventh block of cheese?

6 Indra weighs nine mobile phones. Here are the weights in grams.

    67, 87, 82, 85, 95, 90, 125, 67, 67

  (a) Work out the modal weight.

  (b) Work out the mean weight.

Another mobile phone was added. The mean weight increased by 2 grams.

  (c) What was the weight of the tenth mobile phone?

7 Yousaf calculates the average number of megapixels for 10 digital cameras. Here are his results.

    5 Nikon cameras with 6 megapixels each
    3 Canon cameras with 8 megapixels each
    2 Fuji cameras with 9 megapixels each

Work out the mode, median, mean and range of the data for the 10 cameras.

Exercise 20.2   **Link 20E**

1 Find the mode, median and mean in each of the following and make a comment about each.

  (a) The number of stamps on each of eight letters.
    1, 3, 1, 4, 1, 1, 1, 4

  (b) The number of people in each of seven queues at a supermarket.
    3, 5, 14, 4, 6, 3, 7

  (c) The number of babies born to each of ten women.
    2, 2, 1, 3, 5, 0, 2, 0, 2, 1

  (d) The number of tables booked each night in a restaurant one week.
    3, 3, 3, 7, 12, 12, 2

2 Zep bought 5 boxes of apples at £10 a box, 2 boxes of oranges at £15 a box and 1 box of pineapples at £30 a box.

> 5 boxes of apples cost
> 5 × £10 = £50

(a) Work out the mean cost of a box of fruit.

(b) Which average would Zep use if he wanted to complain about how expensive the fruit was? Why?

(c) Which average would Zep use if he was boasting to his friends what a good deal he had made? Why?

3 The weekly wages for the local football club were

1 manager	£800
1 trainer	£450
11 players	£400 each
4 reserves	£350 each
2 maintenance	£200 each
1 secretary	£250

(a) Calculate the total weekly wages bill.

(b) What is
(i) the mode (ii) the median
(iii) the mean (iv) the range?

> The total paid to the
> 11 players is
> 11 × £400 = £4400

(c) Comment on your answers to part (b).

Exercise 20.3   Link 20F

1 The times (in minutes) taken to change an exhaust in Mr Fixit's garage were

20, 25, 24, 10, 15, 18, 20, 15,
15, 20, 18, 25, 32, 8, 12, 15,
20, 18, 20, 24, 30, 15, 10, 12,
18, 20, 15, 25, 24, 20, 17, 16

(a) Copy and complete this stem and leaf diagram.

Stem	
0	
1	
2	
3	

Key: 0 | 8 means 8

(b) Work out the range of the data.

(c) Work out the mode and the median.

2 The numbers of pounds lost by members of a slimming club in a four-week period were

    10, 12, 4,   6,   3,   4, 8, 4, 15, 12,
    16, 20, 4, 22, 31, 11, 5, 4,   7, 10

(a) Draw a stem and leaf diagram for this data.

(b) Work out the mode.

(c) Work out the median.

> Use stems 0, 1, 2 and 3.

3 The ages at death of twenty World Heavyweight Boxing champions were

    54 73 81 45 91 66 43 87 80 55
    50 66 59 86 60 38 99 53 68 69

(a) Copy and complete this stem and leaf diagram, including a key.

Stem	
3	
4	
5	
6	
7	
8	
9	

> **Remember:** You must always add a key to a stem and leaf diagram.

(b) Find the mode, median, mean and range of this data and comment on your answers.

Exercise 20.4   **Links 20G, H**

1 Here are the numbers of animals Bill has on his farm.

Animal	Frequency	Angle
Cows	100	200°
Sheep	50	
Hens	20	
Pigs	10	
Total	180	

> To work out the angle for cows use
>
> $\dfrac{100}{180} \times 360° = 200°$

(a) Copy and complete the table to show the angle for each animal.

(b) Draw a pie chart to show the data.

2 Bob has only 36 animals on his farm.

Animal	Frequency	Angle
Cows	4	40°
Sheep	20	
Hens	10	
Pigs	2	
Total	36	

(a) Copy and complete the table to show the angle for each animal.

(b) Draw a pie chart to show the data.

3 The numbers of people in the different armed forces in the UK, to the nearest thousand, are shown in the table.

	Number (to nearest thousand)	Angle
Army	115	197°
Navy	42	
Air Force	53	
Total	210	

(a) Copy and complete the table by calculating the angles needed to draw the pie chart.

(b) Draw the pie chart.

4 The table gives data about the contents of a bag of crisps.

Contents	Amount
Carbohydrate	12.5 g
Protein	7.5 g
Sodium	2.5 g
Other	2.5 g
Total	25 g

To work out the angle for carbohydrate use

$$\frac{12.5}{25} \times 360° = 180°$$

Draw a pie chart to represent the data in the table.

5 The pie chart shows data about 120 people on a cruise.

Copy and complete the following table.

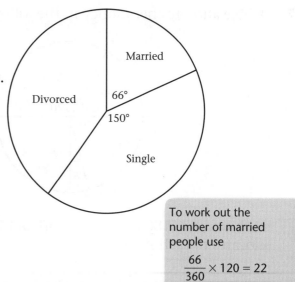

	Angle	Number of people
Married	66°	
Single	150°	
Divorced		
	Total	120

To work out the number of married people use

$$\frac{66}{360} \times 120 = 22$$

6 Sandy asked 90 people who stayed for a school dinner what they ate. Here are the results.

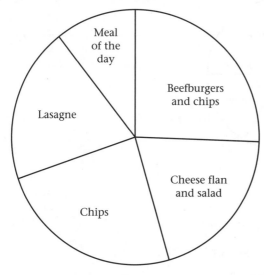

Complete the following table to work out how many students ate each type of meal. You will need a protractor.

Meal	Number of degrees	Number of students
Meal of the day		
Beefburgers and chips		
Cheese flan and salad		
Chips		
Lasagne		
Total	360°	90

7 Debbie allocates her money in the following way.

Debbie earns £180 per week. Work out how much she allocates to each item.

8 Tasmin collects data on favourite sweets. Here is the pie chart she produces to show the data.

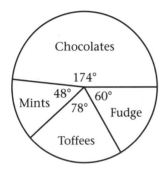

Tasmin interviewed 60 people.

(a) How many liked fudge?

(b) Which was the most popular sweet?

(c) Which was the least popular sweet?

(d) How many people liked toffees?

(e) How many more people liked chocolates than toffees?

21 Formulae and inequalities

Exercise 21.1

Links 21A–C

1 To work out her pay Jessica uses the word formula

Pay = Rate of pay × hours worked + bonus

Work out her pay when she works for 25 hours at a rate of pay of £5 an hour and earns a bonus of £8.

2 To work out his pay Pulin uses the word formula

Pay = Rate of pay × hours worked + bonus

Work out his pay when he works for 40 hours at a rate of pay of £4.50 an hour and earns a bonus of £12.

3 Use the formula

Cost of golf balls = Cost of one golf ball × number of golf balls

to work out the cost of 12 golf balls if one golf ball costs £1.60.

4 Write a word formula to help solve each of these problems. Use your formula to work out the answers.

(a) Trevor works for 35 hours at a rate of pay of £6 an hour. How much should he get paid?

(b) Arti works for 30 hours at a rate of pay of £7.50 an hour. How much should she get paid?

(c) Peter buys 15 bars of chocolate at 60p each. Work out the total cost of the bars of chocolate.

(d) Clara buys 20 newspapers at 25p each. Work out the total cost of the newspapers.

(e) Lucy sold 30 rolls at 45p each. How much money did she collect?

(f) Clive sold 60 roses at £3.50 each. How much money did he collect?

(g) Christine cooked 64 biscuits for a party. Everybody at the party had 2 biscuits. How many people were at the party?

(h) A window cleaner charges £5.60 to clean 7 windows. He charges the same amount to clean each window. What does he charge to clean 1 window?

5 Write an algebraic formula for each part of question 4.

Exercise 21.2

Links 21D–F

1 In this question $a = 2$, $b = 3$, $c = 4$ and $d = 0$.
Work out the value of

(a) $a + b$ (b) $b + c$ (c) $a + b + c$

(d) $3a$ (e) abc (f) abd

(g) $5a - 2b$ (h) $4c + 2a$ (i) $3b - 2a$

(j) $a + 3b - 2c$ (k) $3c - 2d$ (l) $3a + 4b - 4c$

(m) $2ab$ (n) $ac - ab$ (o) $4c - 3a$

(p) $3b - a$ (q) $7c - 5b + a$ (r) $5abc$

> **Remember:**
> $2a$ means $2 \times a$
>
> $\dfrac{a}{2}$ means $a \div 2$
>
> $2a + 1$ means $2 \times a$ then $+ 1$
>
> $\dfrac{a}{2} - 3$ means $a \div 2$ then $- 3$
>
> $2a + 3b$ means work out the values of $2a$ and $3b$ then add these values together.

2 The formula for the perimeter of a square is $P = 4s$.
Find the value of P when

(a) $s = 2$ (b) $s = 5$

(c) $s = 7$ (d) $s = 3.6$

(e) $s = 4.8$ (f) $s = 16$

3 The formula for the volume of a cuboid is $V = lwh$.
Find the value of V when

(a) $l = 2$, $w = 3$, $h = 1$ (b) $l = 6$, $w = 3$, $h = 2$

(c) $l = 8$, $w = 4$, $h = 3$ (d) $l = 12$, $w = 3$, $h = 4$

4 Use the formula $v = u + at$ to work out v when

(a) $u = 2$, $a = 3$, $t = 1$ (b) $u = 4$, $a = 4$, $t = 5$

(c) $u = 10$, $a = 8$, $t = 9$ (d) $u = 8.2$, $a = 10.2$, $t = 5$

Exercise 21.3

Links 21G–K

1 Work out

(a) $7 - 4$ (b) $5 - 8$ (c) $-8 - 5$

(d) $4 + (-5)$ (e) $-5 - (-3)$ (f) $8 - (-4)$

(g) $5 - (-6)$ (h) $10 + (-12)$

2 In this question $a = -2$, $b = 3$, $c = -4$, $d = 0$.
Work out the value of

(a) $a + c$ (b) $c + b$ (c) $d - c$

(d) $c - d$ (e) $c - a$ (f) $a + b$

(g) $b + a$ (h) $c + b + a$ (i) $c + b - a$

(j) $a + b - c + d$ (k) $a - c + b$ (l) $b - c - a$

(m) $d - c + a - b$ (n) $b - a - b$ (o) $c - a + b$

3 Work out

(a) 3×-2 (b) -2×-2 (c) -7×5

(d) 4×-8 (e) -5×-5 (f) -8×4

(g) 3×-9 (h) -6×-4

$+ \times + = +$
$+ \times - = -$
$- \times + = -$
$- \times - = +$

4 In this question let $a = -4$, $b = 3$, $c = -2$ and $d = 0$.
Work out the value of

(a) $a + c$ (b) $b + c$ (c) $a + b + c$

(d) $3a$ (e) ac (f) ab

(g) abc (h) $5a + 3b$ (i) $6b - 2c$

(j) $db + bc$ (k) $ac + ab - bc$ (l) $4ac + 2bc$

5 Use the formula $v = u + at$ to work out v when

(a) $u = 2$, $a = -3$, $t = 3$ (b) $u = -3$, $a = 9$, $t = 5$

(c) $u = 5$, $a = -4.5$, $t = 4$ (d) $u = -10$, $a = 2$, $t = 6$

6 Use the formula $m = np + q(r - s)$ to work out m when

(a) $n = 2$, $p = -3$, $q = 2$, $r = 5$, $s = -2$

(b) $n = -3$, $p = -4$, $q = -5$, $r = 4$, $s = 3$

Exercise 21.4

Link 21L

1 Draw this table of values in your book.

x	-3	-2	-1	0	1	2	3
$y = 2x^2 + 3$							

Remember:
a^2 means $a \times a$
$2a^2$ means $a \times a$ then $\times 2$
$2a^2 + 1$ means $a \times a$ then $\times 2$ then $+ 1$
$a^2 + b^2$ means work out the values of $a \times a$ and $b \times b$ separately then add the values.

Complete the table by substituting the values of x into the formula to find the values of y.

2 If $a = 3$, $b = -2$ and $c = 5$ work out the values of

(a) $a^2 + c$ (b) $c^2 - a$

(c) $a^2 - b$ (d) $a^2 + (a^2 + c)$

(e) $2(b^2 + c^2)$ (f) $3(a + c)^2$

(g) $(a + c)^2 + (a + b)^2$ (h) $3(a + c)^2 - 2(c + b)^2$

(i) $\sqrt{(c^2 + 3ab^2)}$

Remember BIDMAS.
Indices are worked out before +.

Exercise 21.5 Link 21M

1 Work out the value of these algebraic expressions using the values given.

(a) $5a + 3$ if $a = 4$ (b) $4b - c$ if $b = 2, c = 5$

(c) $3p - 2q$ if $p = 5, q = 2$ (d) $xy - z$ if $x = 2, y = 4, z = 3$

(e) $12 + 5t$ if $t = -2$ (f) $p - 3t$ if $p = 4, t = -2$

(g) $4y + 7$ if $y = 4\frac{1}{2}$ (h) $6st$ if $s = \frac{1}{2}, t = \frac{3}{4}$

(i) $4(a + b)$ if $a = 2, b = 3$ (j) $5(x - y)$ if $x = 7, y = 4$

(k) $x(6 - y)$ if $x = 3, y = 2$ (l) $3(8 - t)$ if $t = -2$

(m) $\frac{1}{2}(a + b)$ if $a = 3, b = 5$ (n) $-2(3t + 1)$ if $t = -2$

(o) $3(2x - y)$ if $x = -1, y = -3$ (p) $4(p + 2q)$ if $p = 1, q = -2\frac{1}{2}$

(q) $6(a + b)$ if $a = 2, b = -2$ (r) $\frac{3}{4}(f + g)$ if $f = 2, g = -10$

2 Work out the value of each of these expressions for the given values of the letters.

(a) $\dfrac{x}{5} + 2$ if $x = 15$ (b) $\dfrac{p}{q} - 2$ if $p = 36, q = 3$

(c) $\dfrac{n - 3}{5}$ if $n = 13$ (d) $\dfrac{a + 2b}{3}$ if $a = 1, b = 4$

(e) $\dfrac{ab + c}{2}$ if $a = 2, b = 3, c = 4$ (f) $\dfrac{2pq}{12} - 3$ if $p = 4, q = 6$

(g) $\dfrac{4xy - z}{7}$ if $x = 2, y = 3, z = -4$ (h) $\dfrac{3t}{15} + s$ if $t = 5, s = -1$

(i) $2t - \dfrac{3r}{4}$ if $t = 5, r = 8$ (j) $a - \dfrac{4b}{6}$ if $a = -2, b = -12$

Exercise 21.6 Links 21N, O

1 Use the information in these diagrams to find the values of the letters.

(a)   (b)

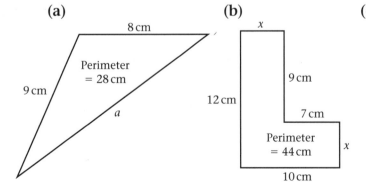

(c)

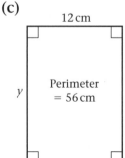

(d)

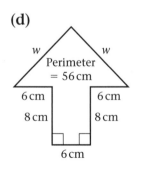

2 Joanna thought of a number.
She multiplied it by 3 then added 5.
The answer was 17.
What number did she think of?

3 Scott thought of a number.
He took away 6 from it, then he multiplied the
answer by 5 to get a final answer of 15.
What number did Scott think of first?

4 Catherine thought of a number, added 9 to it then
multiplied the answer by 4.
This gave an answer of 44.
What number did Catherine think of?

5 Seamus thought of a number, took 8 away and then
divided his answer by 2.
The final answer was 16.
What number did he think of?

6 In each part
   (i) use the diagram to write down an equation in x
   (ii) solve your equation.

(a)

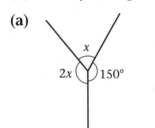

$2x$ $\quad$ $150°$
x

(b)

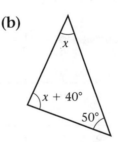

x
$x + 40°$
$50°$

(c)

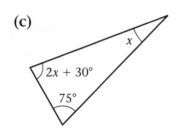

x
$2x + 30°$
$75°$

(d)

$x + 2$

4 $\quad$ Area = 28

(e)

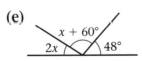

$x + 60°$
$2x$ $\quad$ $48°$

(f)

$x - 4$

5 $\quad$ Area = 15

Remember:

The angles at a point add
up to 360°.

The angles on a straight
line add up to 180°.

The angles in a triangle
add up to 180°.

Exercise 21.7

Links 21P, Q

1 Put the correct sign between these pairs of numbers to make a true statement.

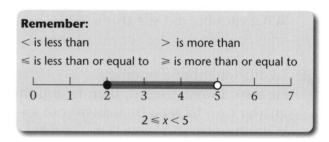

Remember:

$<$ is less than $>$ is more than

$\leqslant$ is less than or equal to $\geqslant$ is more than or equal to

$2 \leqslant x < 5$

(a) 5, 7   (b) 7, 3

(c) 5, 5   (d) 0.2, 0.4

(e) 0.1, 0   (f) 3.16, 3.15

2 Write down the values of x that are whole numbers and satisfy these inequalities.

(a) $1 < x < 4$ (b) $5 < x \leqslant 10$

(c) $1 \leqslant x < 5$ (d) $0 < x < 7$

(e) $3 \leqslant x \leqslant 6$ (f) $5 \leqslant x < 9$

3 Draw number lines from 0 to 10. Shade in these inequalities.

(a) $x > 4$ (b) $x \leqslant 7$

(c) $3 < x < 6$ (d) $4 < x \leqslant 9$

(e) $1 \leqslant x < 4$ (f) $5 < x \leqslant 8$

(g) $8 \leqslant x < 10$ (h) $4 \leqslant x \leqslant 6$

4 Write down the inequalities represented by the shading of these number lines:

(a)

(b)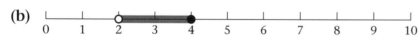

(c)

22 Transformations

1 Draw each shape on squared paper and translate it by the amount shown.

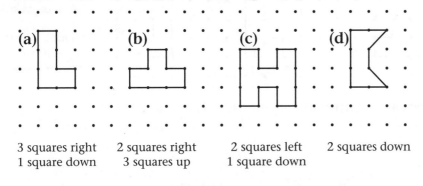

(a)   (b)   (c)   (d)

3 squares right   2 squares right   2 squares left   2 squares down
1 square down   3 squares up   1 square down

A translation is a slide. There is no turning or twisting.

2 Copy the shapes on to squared paper and draw the whole image to match the point that has been translated.

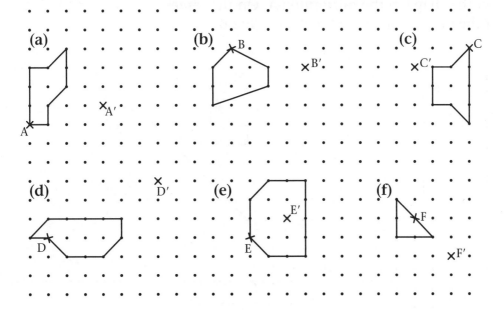

3 Describe the translations for question **2**.

Exercise 22.2   **Link 22B**

1   Draw separate images for each shape after a rotation of 90°
    anticlockwise for each of the centres marked.

> Use tracing paper.

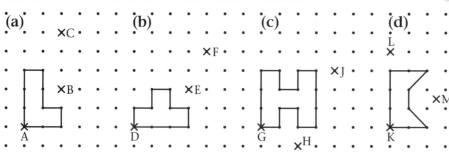

> **Remember:**
> Clockwise is the direction the hands on a clock turn. Anticlockwise is the other way.

2   Using the same shapes and centres as for question **1**, rotate
    by a half turn.

Exercise 22.3   **Link 22C**

For the following questions, copy each diagram on to squared
paper and reflect the shape in the mirror line A. On the same
diagram, reflect the original shape in the mirror line B.

1

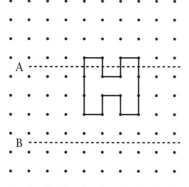

2

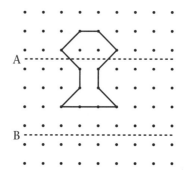

3

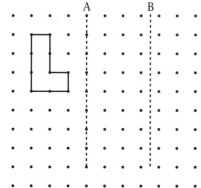

4

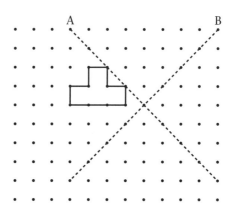

5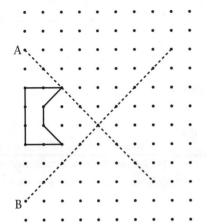

6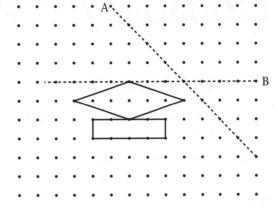

Exercise 22.4

Link 22D

1 In the diagram each shape has two possible centres of enlargement marked. Copy each diagram and show both enlargements. Use a scale factor of 2 for parts **(a)** and **(c)** and a scale factor of 3 for part **(b)**.

> If the scale factor is 2, all the image points are twice as far away from the fixed centre as the object points.

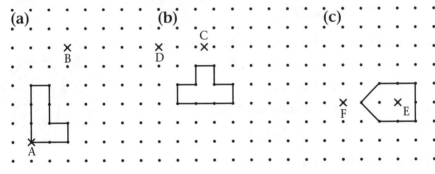

2 Each shape in the diagram has two possible centres of enlargement marked by two letters. Copy each diagram and show both enlargements. Use a scale factor of 2 for the first letter and a scale factor of 3 for the second letter.

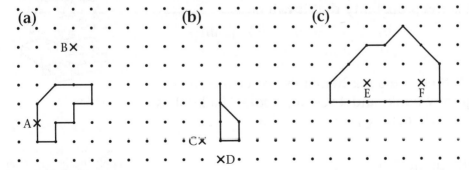

23 Probability

Exercise 23.1
Links 23A, B

1 Draw a 0 to 1 probability scale and mark on it the probability that

> **Remember:**
> Probability = $\dfrac{\text{number of successful outcomes}}{\text{total number of possible outcomes}}$

   (a) Easter Monday follows Easter Sunday
   (b) you will listen to the radio today
   (c) the temperature will not rise next month
   (d) if you drop a table mat it will land face up
   (e) you will see a robin today.

2 A fair six-sided dice is rolled. What is the probability of getting

   (a) a 3   (b) a 4 or 5
   (c) a number less than 3   (d) a 0
   (e) 1 or 6   (f) greater than 5
   (g) 1 or 2 or 4   (h) a 7?

3 The fair spinner is spun. What is the probability of getting

   (a) a 2   (b) an odd number
   (c) 3 or 8   (d) 7 or 4 or 5
   (e) an even number   (f) a prime number
   (g) greater than 4   (h) less than 3?

4 A letter is chosen at random from the word MISSISSIPPI. What is the probability that it is

   (a) an I   (b) an S   (c) a P
   (d) a consonant   (e) an M?

5 A box of replacement lights for a Christmas set has 5 plain, 8 red, 3 blue and 4 green lights. If one is selected at random what is the probability that it will be

   (a) red   (b) blue   (c) plain
   (d) green   (e) yellow?

6 On a bookshelf there are 6 mathematics books, 3 history books, 2 geography books and 4 science books. If a book is taken at random what is the probability that it will be

   (a) a history book
   (b) not a mathematics book
   (c) a science or a mathematics book?

> 'Not a mathematics book' means any other book.

7 Anne is asked to think of a number between 1 and 100 inclusive. Assuming her choice is random, what is the probability that her choice is

(a) 10 or less   (b) higher than 80

(c) an even number   (d) a square number

(e) 23   (f) a cube number?

Exercise 23.2   Links 23C, D

1 Write down the probability of each of the following.

(a) A tree will grow to be 5 km tall.

(b) You can hold your breath for 10 minutes.

(c) If you toss a coin it will land on its edge.

(d) Your pet will live to be 250 years old.

(e) If you roll a dice it will show an even number.

2 Write two statements for each of the following.

(a) A probability of 0.

(b) A probability of 1.

(c) A probability of about $\frac{1}{2}$.

(d) A probability of about $\frac{1}{4}$.

3 A box of chocolates contains 4 soft centres, 3 nuts, 1 plain chocolate and 2 toffees. Freda takes one chocolate without looking. Write (i) as a fraction (ii) as a decimal and (iii) as a percentage the probability that she takes

(a) a soft centre   (b) a toffee

(c) a nut   (d) a plain chocolate.

4 Here are the nets of two different dice.

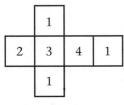 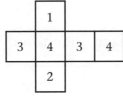

            Die A   Die B

If both dice are rolled what is the probability that
(i) using dice A (ii) using dice B you will get

(a) a 1   (b) a 4

(c) an even number   (d) an odd number

(e) a prime number   (f) a square number

(g) a number greater than 1   (h) a number less than 3?

5 Rashid spins the spinner. Each number is equally likely. What is the probability he will

   (a) get a 1

   (b) *not* get a 1

   (c) get an even number

   (d) *not* get an even number

   (e) get a number greater than 2

   (f) *not* get a number greater than 2?

Exercise 23.3   Link 23E

1 Toss a coin 50 times. Record your results in a table like this.

Coin	Tally	Frequency	Probability
Head			$\overline{50}$
Tail			$\overline{50}$
		Total	

Use your table to answer these questions.

   (a) What is the probability of tossing a head?

   (b) What is the probability of tossing a tail?

   (c) What is the total of the probabilities?

   (d) Explain your result to part **(c)**.

2 Toss a coin another 50 times and complete a table as in question **1**. Compare your results with question **1** and make comments.

3 Toss two different coins 50 times. Record your results in a table like this.

Coins	Tally	Frequency	Probability
HH			
HT or TH			
TT			

   (a) What is the probability of HH?

   (b) What is the probability of HT or TH?

   (c) What is the probability of TT?

   (d) What is the total of the probabilities?

   (e) What is the probability you will not get TT?

Exercise 23.4

Link 23F

1 This sample space diagram shows the outcomes when a coin is tossed and a dice is rolled.
Find the probability of getting

   (a) a head and a 1

   (b) a head and an odd number

   (c) a tail and a 4

   (d) tail or head and a 3

   (e) tail or head and an even number

   (f) a tail and a prime number

   (g) tail or head and a prime number.

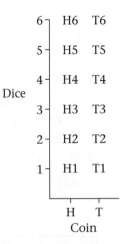

2 The spinner is spun once and the colour recorded.
The spinner is spun a second time and the colour recorded.

   (a) Draw a sample space diagram to show all the outcomes.

   (b) List all the outcomes.

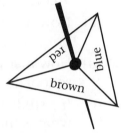

3 From a normal pack of playing cards a card is chosen, its suit recorded and then it is replaced. A second card is chosen, its suit recorded and then it is replaced.

   (a) Draw a sample space diagram to show all the outcomes.

   (b) List all the outcomes.

Remember: There are 4 equal suits in a pack of 52 cards: clubs, hearts, diamonds, spades.

4 John can either go straight on or turn left at a junction. List all the possible outcomes after he has passed two junctions.

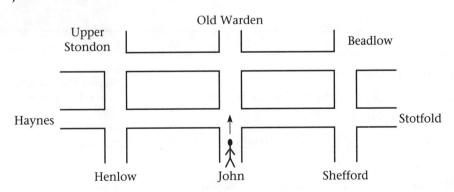

Exercise 23.5

1 The numbers of students that used the computer room before
   school, at break and at lunchtime are shown in this table. A
   student can only use the computer room once a week.

	Monday	Tuesday	Wednesday	Thursday	Friday	Totals
Before school	18	12		9	14	69
Break	33	25	28	22	31	
Lunchtime		41	37	34	43	
Total	85		81			

(a) Copy and complete the table.

(b) How many students used the computer room on
   Wednesday before school?

(c) How many students used the computer room on
   Thursday?

(d) How many students used the computer room before
   school during the week?

(e) If a student from the above table is chosen at random
   what is the probability that they used the computer
   room during

   (i)   Monday break   (ii) Wednesday lunchtime

   (iii) Friday   (iv) lunchtime in the week?

2 Mary and William asked 100 customers how they travelled to
   Edinburgh and in which month they travelled. Some of the
   information is recorded in this table.

	Coach	Rail	Car	Air	Totals
January	6		17	3	30
February		11	7	5	45
March					
Total		28	32	11	100

(a) Copy and complete the table.

(b) A customer is selected at random. What is the
   probability they travelled

   (i)   in February   (ii) in January or March

   (iii) by car   (iv) by car or air

   (v)   not by coach   (vi) by rail in March

   (vii) by air in March   (viii) in January not by rail?

> **Remember:** A random
> selection is one in which
> each person has the
> same chance of being
> chosen.

3 You need 10 drawing pins. Throw them on to a flat surface. Record the number that land with points uppermost. Repeat this 20 times. Use your results to obtain an estimate for the probability of a drawing pin landing on its head.

4 Ramón tests the quality of microchips in a factory. He takes 6 samples, each of 20 microchips, and tests them for quality. They either pass or fail. The table shows the number of microchips in each sample that fail.

First 20 microchips	Second 20 microchips	Third 20 microchips	Fourth 20 microchips	Fifth 20 microchips	Sixth 20 microchips
2 fail	3 fail	1 fails	3 fail	1 fails	2 fail

Ramón then works out the proportions of microchips that fail throughout his tests. The table shows some of his work.

	Number of fails	Proportion of fails
First 20 microchips	2	$\frac{2}{20} = 0.1$
Second 20 microchips	2 + 3 = 5	$\frac{5}{40} = 0.125$
Third 20 microchips		
Fourth 20 microchips		
Fifth 20 microchips		
Sixth 20 microchips		

(a) Copy and complete the table.

(b) Illustrate your results using a line graph.

(c) Comment on what happens to the proportion of fails as the number of tests increases.

24 Presenting and analysing data 2

1 Students' marks in a maths paper and a science paper were

Maths	25	37	69	43	80	74	56	29	48	59	62
Science	31	38	66	47	76	79	58	30	54	56	65

  (a) Draw a scatter diagram to represent this data.
  (b) What type of correlation do you find?

2 What type of correlation, if any, would you expect if you compared the following data? Explain your answers.
  (a) Shoe sizes and heights of people.
  (b) The size of cars' engines and their highest speeds.
  (c) The number of bedrooms in houses and the number of people living in them.
  (d) The length of people's toe nails and their age.
  (e) Students' test scores in French and English.
  (f) A person's income and their number of pets.

3 The handspan and the length of their little finger were measured for 10 students.

Handspan (mm)	201	182	174	207	203	191	197	184	189	171
Little finger (mm)	64	58	51	68	70	59	63	59	62	56

  (a) Draw a scatter diagram to represent this data.
  (b) What type of correlation do you find?
  (c) Draw and label the line of best fit on your scatter diagram.
  (d) Use your line of best fit to predict the little finger length for a person whose handspan is 190 mm.

4 The height and shoe size of 12 people were measured.

Height	170	167	177	174	169	161	172	166	180	158	182	179
Shoe size	7	8	10	9	8	7	8	6	12	5	11	10

  (a) Draw a scatter diagram to represent this data.
  (b) What type of correlation do you find?
  (c) Draw and label the line of best fit on your scatter diagram.
  (d) Use your line of best fit to estimate the height of a person whose shoe size is $7\frac{1}{2}$.

5 The table shows details of attendance and entrance charges for visitors to the properties of a national organisation.

(a) Draw a scatter diagram and describe the correlation.

(b) Give possible reasons for your answer.

Attendance (thousands)	Charge (£)	Attendance (thousands)	Charge (£)
297	4.20	211	6.00
200	6.00	188	3.00
180	3.50	158	5.00
141	5.50	114	5.00
108	4.90	96	2.30
81	3.60	82	5.40
67	4.00	50	3.30

6 Joe has 17 cars for sale. The tables shows the ages and prices of the cars.

Age in years	5.5	5	4	5	4	3	4	3.5
Price in €	1250	2000	1500	2500	2000	2500	3000	3750

Age in years	3	3	2	1	3	4	1	5	4
Price in €	2000	3500	3250	4000	3000	2500	3750	1500	2250

(a) Draw a scatter diagram.

(b) Describe the correlation between the ages of these seventeen cars and their prices.   [E]

Exercise 24.2   Links 24B, C

1 Find the mean and median of the following sets of data.

(a) 2, 9, 1, 16, 11, 8, 3

(b) 22, 29, 21, 36, 31, 28, 23

(c) 6, 27, 3, 48, 33, 24, 9

(d) 2.6, 9.6, 1.6, 16.6, 11.6, 8.6, 3.6

> **Remember:** The mean is the value that occurs most often.
>
> The median is the middle value, when the data is arranged in order of size.

2 Use the information given to find the value of n in each of the following (where n is a missing number in the list).

(a) 4, 12, 7, n, 13   the mean is 10

(b) 7, 4, 2, 2, 7, 3, n the mode is 2

(c) 15, n, 9, 11, 9, 4, 5   the mean is 9

(d) 12, 16, 9, 11, n, 5   the range is 16

(e) 4, 5, 4, 5, n, 5, 4, 4   the modes are 4 and 5

(f) n, 15, 21, 7, 5, 1, 9, 19   the median is 10

(g) 32, 43, 9, 15, n, 12   the median is 18

(h) 36, 48, 96, 27, n, 72   the range is 81

3 The average rainfall in Braintree for the past five years has been 48.7 cm.
How much rain must fall this year to bring the mean for the six years up to 50 cm?

> The amount of rain in six years is
> $6 \times 50 = 300$ cm

4 The scores for a basketball team are 72, 84, 102, 86, 73, 90 and 83.
The scores in the next two matches bring the mean up to 90.
What was the total score for these two games?

5 A local government records the number of children in each of 30 families. The results are shown in the table.

Number of children	Frequency	Frequency × number of children
0	3	$0 \times 3 = 0$
1	4	$4 \times 1 = 4$
2	10	$10 \times 2 = 20$
3	6	
4	4	
5	2	
6	1	
Total	30	

(a) Copy and complete the table.

(b) Work out the mean number of children in these families.

6 In a fitness test, thirty 45-year-olds were asked to run to the top of a flight of stairs. Their times, in seconds, were recorded and are summarised in the table.

Using the data in the table, work out the

(a) mode

(b) median

(c) mean

(d) range.

Time (seconds)	Frequency	Frequency × time
10	1	
11	2	
12	7	
13	10	
14	6	
15	3	
16	1	
Total	30	

7 The numbers of photos taken on '36-exposure' films are listed in the table below.

Number of photos	34	35	36	37	38
Frequency	1	3	26	31	4

Work out **(a)** the median **(b)** the mean.

8 The Underground fares in London are set according to the zones travelled through. The table shows the fares payable when travelling from Zone 1 to other zones and the number of passengers who buy these tickets in one hour.

	Zone 1	Zone 2	Zones 3 & 4	Zones 5 & 6
Fare	£2.00	£2.30	£2.80	£3.80
Frequency	103	72	23	12

Work out **(a)** the mode **(b)** the mean.

9 A manager recorded the number of absences for the employees in his department in a year. Here are his results.

Number of absences	Frequency	Frequency × number of absences
0	2	
2	5	
4	10	
6	4	
8	5	
10	2	
12	1	
30	2	
Total		

(a) Using the data in the table, work out the
   (i) mode **(ii)** median **(iii)** mean **(iv)** range.
(b) Which average would the manager use if he wanted to complain about the absence rate in his department? Why?
(c) Which average would the manager use if he wanted to boast about his department's work record? Why?

10 The table gives the weekly wages earned by workers in a small factory.

Job	Number of workers	Weekly wage
Machine operator	10	£200
Charge-hand	5	£250
Foreman	3	£300
Production manager	1	£500
Factory owner	1	£1000

(a) Use the data in the table to work out the mode, median and mean weekly wage.

(b) Which average would the factory owner use if he was trying to avoid paying his workers more money? Why?

(c) Which average would the machine operators use if they wanted to ask for more money? Why?

Exercise 24.3   Links 24D, E

1 Work out the (i) range, (ii) interquartile range for the following sets of data.

(a) 5, 9, 10, 12, 12, 16, 19

(b) 3, 8, 16, 11, 9, 3, 7, 13, 3, 5, 0

(c) 15, 8, 0, 27, 100, 3, 4, 7, 15, 25, 7, 6, 9, 9, 30

2 Work out the (i) range, (ii) interquartile range for the data shown in each frequency table.

(a) The numbers of advertising e-mails received by 23 people in an office on a particular day.

Number of e-mails	0	1	2	3	4
Frequency	1	3	3	7	9

Remember:
The lower quartile is the
$\left(\dfrac{23 + 1}{4}\right)$ = 6th value.
The upper quartile is the
$\dfrac{3(23 + 1)}{4}$ = 18th value.

(b) The number of years married before the birth of a baby.

Years	0	1	2	3	4	5	6	7
Frequency	21	15	6	7	5	1	3	1

(c) The number of calls to the emergency services each day in a year.

Number of calls	1	2	3	4	5	6	7	8	9	10
Frequency	9	37	55	88	57	30	27	22	18	2

3 A biologist is studying the population of a particular species of snail. The numbers of snails found in 10 m square sections of a field are summarised in the table.

Number of snails	0–2	3–5	6–8	9–11	12–14
Frequency	8	13	10	6	3

Use the middle value of the class intervals: 1, 4, 7…

Work out an estimate for the mean number of snails in each 10 m square section.

4 Polly uses a metal detector to find lost coins on a beach. The numbers of lost coins she finds on 25 days are summarised in the table.

Number of coins	0–4	5–9	10–14	15–19
Frequency	12	7	5	1

(a) What is the modal class?

(b) Work out an estimate for the mean. Why is this 'an estimate'?

5 The table gives data about the heights of 100 international netball players.

Height, x (inches)	Frequency, f	Middle value	$f \times x$
$66 \leqslant x < 68$	7	67	$7 \times 67 = 469$
$68 \leqslant x < 70$	27		
$70 \leqslant x < 72$	36		
$72 \leqslant x < 74$	25		
$74 \leqslant x < 76$	5		

(a) Copy and complete the table.

(b) What is the modal class interval?

(c) What is the class interval in which the median lies?

(d) Work out an estimate for the mean height of these netball players.

6 The table gives data about the CO_2 emissions, in grams per kilometre (g/km), for a particular make of car.

(a) Work out an estimate for the mean.

The target amount of CO_2 emission for this type of car is 178 g/km.

Emission, x	Frequency, f
$155 \leqslant x < 165$	2
$165 \leqslant x < 175$	5
$175 \leqslant x < 185$	3
$185 \leqslant x < 195$	2
$195 \leqslant x < 205$	1

To find the middle value of the interval $155 \leqslant x < 165$ use $\dfrac{155 + 165}{2} = 160$

To meet the target, the emissions should be at or below it.

(b) Have these cars met the target amount of CO_2 emission? Explain your answer.

25 Pythagoras' theorem

1 Find the length of the hypotenuse in these right-angled triangles.

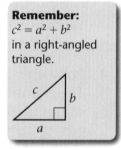
(a)

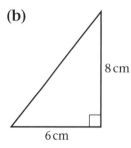

4 cm

3 cm

(b)

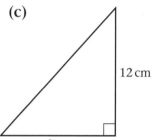

8 cm

6 cm

(c)

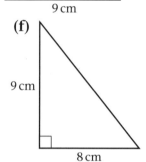

12 cm

9 cm

(d)

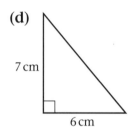

7 cm

6 cm

(e)

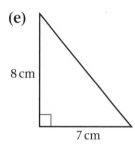

8 cm

7 cm

(f)

9 cm

8 cm

2 Calculate the lengths marked with letters in these triangles.

(a)

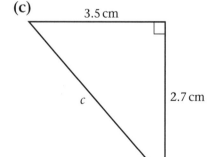

4.5 cm   6.4 cm

a

(b)

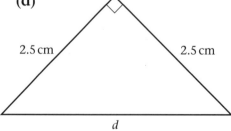

b

3.4 m   4.8 m

(c)

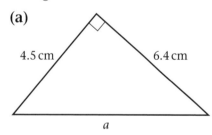

3.5 cm

c

2.7 cm

(d)

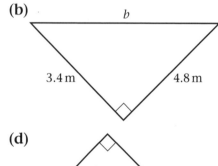

2.5 cm   2.5 cm

d

3 A ladder is resting against the wall of a house.
The foot of the ladder is 3 m from the base of the wall and
the top of the ladder is 4 m from the base of the wall.
How long is the ladder?

4 Calculate the span of the roof
truss shown in the diagram.

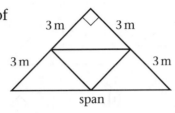

3 m   3 m

3 m   3 m

span

5 Susie is flying her kite on a horizontal playing field.
The string is taut and the kite is 100 m above the ground.
The kite is 300 m from Susie in a horizontal direction.
How long is the kite string?

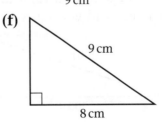

100 m

300 m

Exercise 25.2   Link 25C

1 Calculate the lengths of the unmarked sides in these
right-angled triangles.

> You may need to use a
> calculator.

(a)

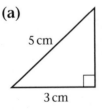

5 cm

3 cm

(b)

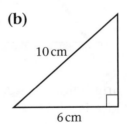

10 cm

6 cm

(c)

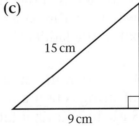

15 cm

9 cm

(d)

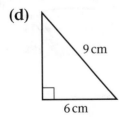

9 cm

6 cm

(e)

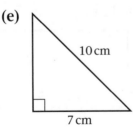

10 cm

7 cm

(f)

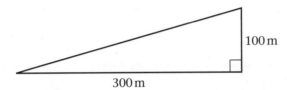

9 cm

8 cm

2 Calculate the lengths marked with letters in these
triangles. Give your answers correct to 1 decimal place.

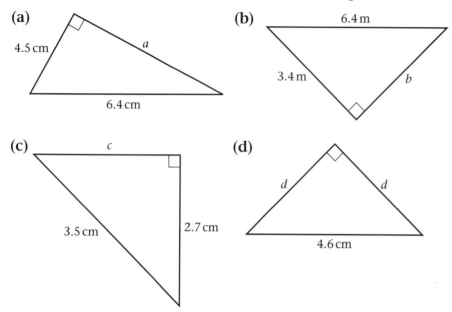

(a) 4.5 cm a 6.4 cm

(b) 6.4 m 3.4 m b

(c) c 3.5 cm 2.7 cm

(d) d d 4.6 cm

3 Keith used his 6-metre long ladder to clean
his upstairs windows. He placed the ladder
2 metres away from the foot of the wall.
How far up the wall did the ladder reach?

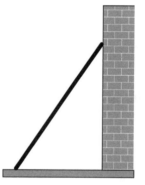

4 Meg used her 8-metre long ladder to
paint her upstairs windows. She placed
the top of the ladder 6 metres above the
ground. How far away from the base of
the wall was the foot of the ladder?

5 Susie is flying her kite on a horizontal playing field.
The 300 m length of string is taut and the kite is
200 m away from Susie in a horizontal direction.
How far from the ground vertically is the kite?

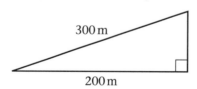

300 m 200 m

26 Advanced perimeter, area and volume

Exercise 26.1   **Link 26A**

1 Find the area of these shapes.

(a)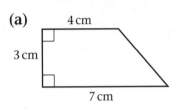

(b)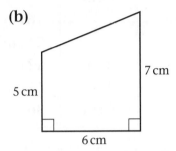

(c)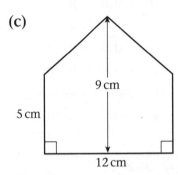

(d)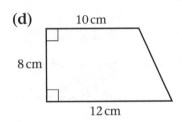

(e)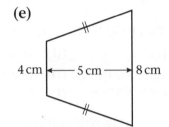

(f)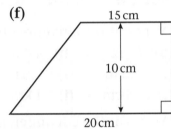

> **Remember:**
> Perimeter is distance around the shape.
> Area is space inside the shape.
> Area of a trapezium $= \frac{1}{2}(a+b)h$, where a and b are the parallel sides.

2 Find the shaded area.

(a)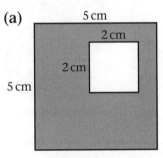

(b)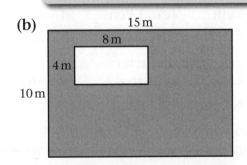

(c)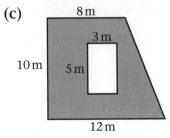

(d)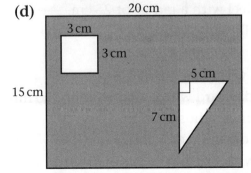

3 Copy this table into your book and complete the missing measurements.

Shape	Base	Vertical height	Area
(a) Parallelogram	6 cm	5 cm	
(b) Parallelogram	8 cm	6 cm	
(c) Parallelogram	8 cm		16 cm^2
(d) Parallelogram	7 cm		21 cm^2
(e) Parallelogram		6 cm	42 cm^2
(f) Parallelogram		2 cm	12 cm^2
(g) Parallelogram		9 cm	36 cm^2

Exercise 26.2

Links 26B–I

1 Work out the circumference of a circle with diameter
(a) 5 cm   (b) 4 cm   (c) 6 cm   (d) 10 cm
(e) 1.5 m   (f) 2.8 m   (g) 3.6 m   (h) 12 m
(i) 25 cm   (j) 1 km   (k) 2.25 m   (l) 5.6 cm

2 Work out the circumference of a circle with radius
(a) 5 m   (b) 3 cm   (c) 2 m   (d) 2.5 m
(e) 3.5 cm   (f) 4.5 m   (g) 2.4 m   (h) 5.6 cm
(i) 20 mm   (j) 30 mm   (k) 2.1 m   (l) 1 m

3 Find the area of a circle with radius
(a) 5 cm   (b) 7 cm   (c) 11 cm   (d) 4 cm
(e) 2.5 cm   (f) 3.2 cm   (g) 5.4 m   (h) 2.9 m

4 Find the area of a circle with diameter
(a) 8 cm   (b) 6 cm   (c) 10 cm   (d) 18 cm
(e) 3.2 cm   (f) 8.4 cm   (g) 6.6 m   (h) 12.4 m

5 A wheelbarrow has a wheel with a diameter of 30 cm.
Work out the circumference of the wheelbarrow wheel.

6 A bicycle has a wheel with a radius of 25 cm.
Work out the circumference of the bicycle wheel.

7 Work out the area of a
(a) circular pond with a radius of 0.9 m
(b) circular birthday card with a diameter of 120 mm
(c) crop circle with a radius of 15 m
(d) pencil with a diameter of 1 cm
(e) circular mill wheel with a radius of 30 cm.

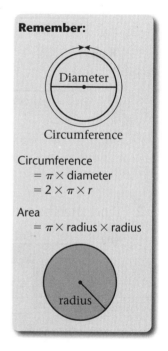

Remember:

Diameter

Circumference

Circumference
= $\pi \times$ diameter
= $2 \times \pi \times r$

Area
= $\pi \times$ radius $\times$ radius

radius

8 Robin travels 600 metres on a bicycle. The circumference of the bicycle wheel is 75 cm.
How many times does the wheel rotate on the journey?

9 Sylvie travels 1.2 kilometres on her bicycle. The diameter of the bicycle wheel is 60 cm.

(a) Work out the circumference of the bicycle wheel.

(b) How many times does the wheel rotate on the journey?

Exercise 26.3   Links 26J, K

1 A circle has an area of 10 cm².
Work out the radius.

2 A tree with a circular trunk has a circumference of 2.8 m.
Work out the diameter of the tree trunk.

3 Copy this table into your book.
Work out the diameter and radius of the circles.

Circumference	Diameter	Radius
(a) 50 cm		
(b) 45 cm		
(c) 10 mm		
(d) 15 cm		
(e) 3.14 m		

4 One of Sven's rollerblade wheels rotates 5000 times when he travels 600 metres.
Work out the diameter of the wheel.

5 The circumference of a circular village pond is 60 metres.
Work out the radius.

6 A circular table top has an area of 3 m².
Work out the diameter of the table.

7 Work out, in terms of π, the area and circumference of a circle with

(a) radius

    (i) 3 cm   (ii) 10 cm   (iii) 20 mm

(b) diameter

    (i) 6 cm   (ii) 5 m   (iii) 10 mm

8 A table top in the shape of a circle has a diameter of 90 cm.

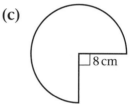

    (a) Work out the circumference of the table top in terms of π.

    (b) Work out the area of the table top in terms of π.

9 The area of a circle is $100\pi\ \text{cm}^2$. Calculate
    (a) the radius
    (b) the circumference of the circle.

Exercise 26.4   **Link 26L**

In this exercise give all your answers to 3 significant figures.

1 Find the area and perimeter of these shapes.

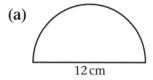
(a)

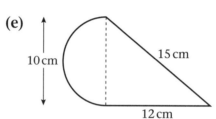
(b)   **(c)**

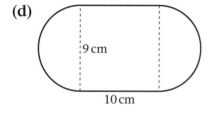

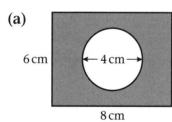
(d)   **(e)**

2 Find the shaded area of each of these shapes.

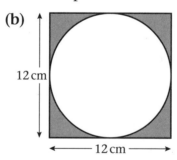
(a)   **(b)**

3 Shane has a lawn in the shape of a rectangle with a semicircle cut out of it.
Calculate the area of the lawn.

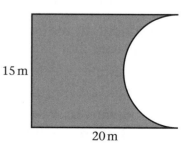

4 Here is a plan of a circular swimming pool.
The pool is surrounded by a path of width 1 m.
The radius of the swimming pool is 5 m.
Calculate the area of the path.

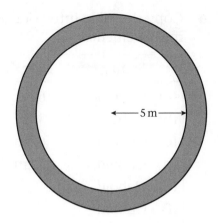

Exercise 26.5 Link 26M

1 Find the volume of each of the following solids.

(a)
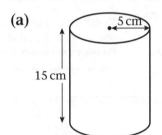

(b)
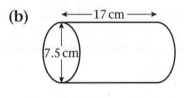

> **Remember:**
> Volume =
> area of end × length
>
>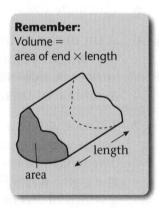

(c)
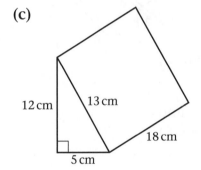

(d)
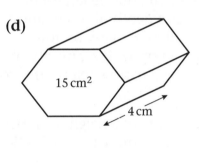

2 How many cubes with edge length 2 cm will fit into a cuboid measuring 10 cm by 8 cm by 6 cm?

3 Work out the volumes of these prisms.

(a)
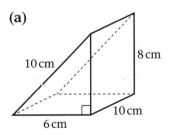

(b)
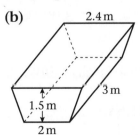

(c)
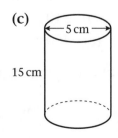

4 Copy the table into your book and fill in the missing measurements.

	Shape	Length	Width	Height	Volume	Surface area
(a)	Cuboid	6 cm	4 cm	2 cm		
(b)	Cube	3 cm				
(c)	Cuboid	6 cm		3 cm	36 cm³	
(d)	Cuboid	8 cm		2 cm	48 cm³	
(e)	Cuboid		4 cm	2 cm	24 cm³	
(f)	Cuboid		5 cm	3 cm	75 cm³	
(g)	Cuboid	4 cm	2 cm		32 cm³	
(h)	Cube					600 cm²

5 Calculate the surface area of the prisms in question **3**.

6 A tin of sweets in the shape of a cylinder with radius 12 cm and height 12 cm has a paper wrapper covering the curved surface. Calculate the area of the paper wrapper. You may assume that there is no overlap.

> Sketch a diagram of the tin to help you. Label the tin with the dimensions.

7 The body of this wheelbarrow has the shape of a prism with a cross-section that is a trapezium. The length of the top is 1.2 m and the length of the bottom is 80 cm. It has a depth of 35 cm and a width of 50 cm. Work out how many barrow-loads it would take to fill a skip with a capacity of 8 m³. (You may assume that a barrow-load is the volume of the barrow.)

8 How many cylindrical glasses with a height of 150 mm and a diameter of 50 mm can be filled from a 3-litre bottle of cola?

Exercise 26.6
Links 26N, O

1 These weights are given to the nearest gram.
Write down the greatest and least weights they could be.

(a) 153 g   (b) 210 g   (c) 3.000 kg

(d) 16 g   (e) 400 g

2 These race times are correct to the nearest $\frac{1}{10}$ second.
Write down the greatest and least time they could be.

(a) 10.2 s   (b) 21.6 s

(c) 41.4 s   (d) 50.0 s

3 State suitable units for measuring

(a) the volume of a bucket

(b) the height of a mountain

(c) the weight of a cucumber

(d) the time to boil an egg

(e) the area of a field

(f) the area of Scotland

(g) the weight of an ocean liner.

Exercise 26.7

Links 26P–S

1 Rose walked 1400 metres in 20 minutes.
What was her average speed?

2 The average speed of a train travelling from London to York is 150 km/h.
The distance is 328 km.
How long does the journey take?

3 Copy and complete the table.

Journey	Distance in km	Journey time	Average speed in km/h
London–Glasgow	642	5 h 22 min	120
Reading–Exeter	219	2 h 03 min	
Birmingham–Liverpool	141		85
York–Newcastle	128	59 min	
Cardiff–Southampton		2 h 29 min	74
Colchester–Norwich	101		93
Plymouth–Penzance	128	1 h 58 min	
Portsmouth–Brighton		1 h 33 min	46
Leeds–Carlisle	181		68
Bristol–Swansea	130	1 h 27 min	
London–York		1 h 51 min	163

4 Copy and complete the table.

Country	Area in km²	Population (1991)	Population density
Nepal	147 000	18.5 million	125.9
Norway	324 000	4.25 million	
Paraguay	407 000	million	10.1
Romania	229 000	22.8 million	
Sierra Leone	71 700	3.52 million	
Sudan	2 510 000	million	8.2
Sweden		8.59 million	20.9
Syria	184 000	9.05 million	
UK		56.5 million	233.3
Barbados	431	259 thousand	
Jamaica	11 000	million	225.4

5 The density of the Earth is 5.5 g/cm³.
The Earth is almost a sphere with radius 6378 km.
Calculate an estimate for the mass of the Earth.

Volume of a sphere $= \frac{4}{3} \times \pi r^3$

6 Change 40 metres per second to kilometres per hour.

7 Change 25 kilometres per hour to metres per second.

8 Change 1800 miles per hour to miles per second.

9 Change 50 centimetres per second to metres per minute.

10 The speed of sound at sea level is 760 miles per hour.
   (a) Change this into kilometres per hour.
   (b) What is the speed of sound at sea level in metres per second?
   (c) Is 20 kilometres per minute faster than the speed of sound?

11 The density of beech wood is 700 kg per m³.
The volume of a beech wood statue is 720 cm³.
Work out the weight of the statue.

12 The density of horse chestnut wood is 540 kg per m³.
   (a) Change this to g per cm³.
   (b) How much does a cuboid of horse chestnut wood measuring 5 cm by 15 cm by 20 cm weigh?

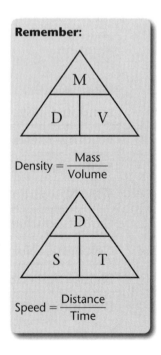

Remember:

Density $= \dfrac{\text{Mass}}{\text{Volume}}$

Speed $= \dfrac{\text{Distance}}{\text{Time}}$

13 A person on a business trip travels from London to Birmingham (184 km), from Birmingham to Nottingham (68 miles) and from Nottingham back to London (126 miles). Work out the total distance travelled in **(a)** miles and **(b)** kilometres.

14 A small gold bar is in the shape of a cuboid. The cuboid is 6 cm wide, 15 cm long and 4 cm high.
Given that 1 cm^3 of gold weighs 19.35 g, work out the weight of the gold bar in kilograms.

15 28.4 grams = 1 ounce
16 ounces = 1 pound
14 pounds = 1 stone

   (a) Find how many grams are equivalent to 1 stone.

   (b) Work out, in pounds and ounces to the nearest ounce, the weight of a baby weighing 3720 g.

16 Roger Bannister was the first man to run a mile in 4 minutes. Work out his average speed in km h^{-1} and metres per second.

27 Describing transformations

1 Copy this grid on to squared paper.
Translate the shape T using the following vectors.

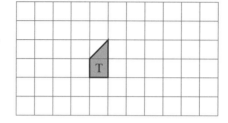

(a) $\begin{pmatrix} 5 \\ 2 \end{pmatrix}$, call it A

(b) $\begin{pmatrix} 5 \\ -2 \end{pmatrix}$, call it B

(c) $\begin{pmatrix} -4 \\ -2 \end{pmatrix}$, call it C

(d) $\begin{pmatrix} -1 \\ 1 \end{pmatrix}$, call it D

> **Remember:** In a translation vector the top value is always the amount moved horizontally and the bottom number is the amount moved vertically.

2 The shape S has been moved into four different positions A, B, C, D.
Write down the translation vectors for each of the four translations.

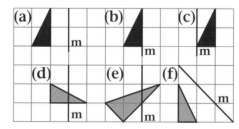

3 Copy this diagram on to squared paper.
For each shape reflect it in the mirror line **m**.

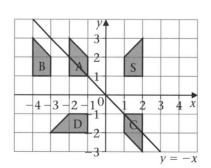

> Remember reflection always happens in a line.

4 Shape S has been reflected four times.
Describe fully each of these reflections that takes S to positions A, B, C, D.

> **Remember:** Rotation is always a movement by an angle or turn about a point. It can be clockwise or anticlockwise.

5 Copy the triangles on to squared paper. Rotate each triangle about the point P marked with a cross, by the angle and direction given.

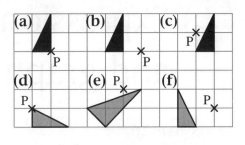

    (a) 90° clockwise   **(b)** 270° anticlockwise

    (c) 180° anticlockwise   **(d)** 90° anticlockwise

    (e) 180° clockwise   **(f)** 45° clockwise

6 Shape S has been rotated four times. Describe fully each of these rotations that takes S to positions A, B, C, D.

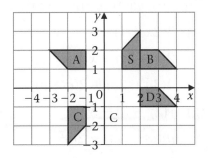

7 Copy this diagram on to squared paper.

Enlarge shape T by the following.

    (a) Scale factor 2 from point P.

    (b) Scale factor 3 from point Q.

    (c) Scale factor 2.5 from point R.

    (d) Scale factor $\frac{1}{2}$ from point R.

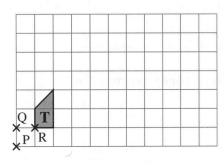

Remember:
An enlargement is always by a scale factor from a point.

8 Describe the enlargement that moves

    (a) shape P on to shape Q

    (b) shape T on to shape R

    (c) shape T on to shape S

    (d) shape R on to shape T.

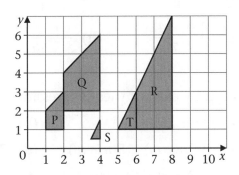

9 Copy this diagram on to squared paper. With P as the centre of enlargement, draw the shaded shape after an enlargement by scale factor $\frac{2}{3}$.

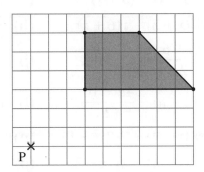

28 Expressions, formulae, equations and graphs

1 Simplify each of these expressions.

(a) $3(2x + 1) + x$

(b) $5(2a + 1) - 3a$

(c) $2(x + 3y) + x + y$

(d) $3(3p - q) + 4(p + 2q)$

(e) $4(3a - 2b) + 2(a + b)$

(f) $x(x - 2) + x$

(g) $y(2y - 3) - 5y$

(h) $5n(2 + 3n) - 6n$

(i) $m(m^2 + m) - 2m^2$

(j) $4t^2(t - 1) - t^3$

(k) $\frac{1}{2}(2x^2 + 4x) + x^2 - x$

(l) $3n^2(n^3 + 2n) + n(n^2 - 3n)$

(m) $5x(2x + x^3) - x^4$

(n) $y^2(3y^2 - 5y) + 2y^3(1 - 3y)$

> Expand the brackets first.
> $3(2x + 1) = 6x + 3$

2 Simplify each of these expressions.

(a) $\dfrac{3n + 12}{3}$

(b) $\dfrac{15x - 10}{5}$

(c) $\dfrac{2a(3a + 6)}{2}$

(d) $\dfrac{6x^2 - 12x}{3}$

(e) $\dfrac{8r + 20s}{4}$

(f) $\dfrac{12a + 6b + 9c}{3}$

(g) $\dfrac{49x - 42y}{7}$

(h) $\dfrac{6a^2 + 4a^2}{5a}$

(i) $\dfrac{5(6t + 8s)}{10}$

3 Simplify each of these expressions.

> **Remember:** Include the signs
> $-4(x - 1) = -4x + 4$
> $(-4 \times -1 = +4)$

(a) $5x - 2(x - y)$

(b) $8t - 3(2t - 1)$

(c) $3(2n + m) - 5(n - m)$

(d) $2(3x - 2) - 4(x - 1)$

(e) $4(3 - x) - (3 + x)$

(f) $c(a - b) - c(a + b)$

(g) $6y(x + 3) - y(6 + x)$

(h) $\frac{1}{2}(4x^2 + 6) - (1 - x^2)$

(i) $\dfrac{4x + 6}{2} - (x - 2)$

(j) $5y - 1 - \dfrac{6y - 2}{2}$

> **Remember:** The line acts like a bracket.

(k) $3(2 - a) - \dfrac{4a - 10}{2}$

(l) $q(p - 1) - q(1 - p)$

(m) $x(3x - 4) - 2(2x^2 - 5)$

(n) $-3(b - a) - (a + b)$

(o) $y(5x + 4) - 2y(x - 3)$

(p) $8r(2 - \frac{1}{2}t) + 3r(1 + t)$

(q) $3x - 3(x - y)$

(r) $4(1 - a^2) - 3(a + a^2)$

Exercise 28.2

Link 28C

1 Multiply out the brackets and simplify where possible.

  (a) $(x + 5)(y + 3)$ **(b)** $(x + 3)(x + 5)$

  (c) $(y + 2)(y + 7)$ **(d)** $(a + 4)(b + 2)$

  (e) $(n + 1)(m + 1)$ **(f)** $(p + 2)(p + 4)$

  (g) $(2a + 1)(a + 1)$ **(h)** $(b + 3)(2b + 4)$

  (i) $(2p + 5)(p + 2)$ **(j)** $(2a + 3)(3b + 2)$

  (k) $(x - 4)(x - 5)$ **(l)** $(y - 2)(y - 5)$

  (m) $(a - 5)(a - 7)$ **(n)** $(2p - 3)(p - 2)$

  (o) $(3q - 5)(4q - 1)$ **(p)** $(x - 1)(x - 2)$

  (q) $(a - 2)(b - 3)$ **(r)** $(3z - 4)(4z - 3)$

  (s) $(5x - 2)(2x - 5)$ **(t)** $(x - 2)(x + 2)$

  (u) $(y - 3)(y + 3)$ **(v)** $(a + 5)(a - 5)$

  (w) $(3x - 2)(x + 7)$ **(x)** $(2a + 1)(a - 4)$

  (y) $(2b + 5)(3b - 2)$ **(z)** $(5n + 7)(3n - 4)$

2 Multiply out the brackets and simplify.

  (a) $(p + 2q)(q + 2p)$ **(b)** $(x + y)(x - y)$

  (c) $(x - y)(x + 2y)$ **(d)** $(n - 2m)(n + 3m)$

  (e) $(a + 3b)(3a - b)$ **(f)** $(2x + 3y)(3x + 4y)$

  (g) $(4a - 1)(a + 2)$ **(h)** $(3x - 2)(4x + 3)$

  (i) $(x + 5)^2$ **(j)** $(y + 1)^2$

  (k) $(2a + 3)^2$ **(l)** $(4x - 3)^2$

  (m) $(2 - x)^2$ **(n)** $(3 - 5y)^2$

  (o) $(a - 3)^2$ **(p)** $(x + y)^2$

  (q) $(3x + 2y)^2$ **(r)** $(4p - 5y)^2$

  (s) $(3x - 1)(x + 4)$ **(t)** $(3b + 2)(5b + 4)$

  (u) $(x - 2y)(x + 2y) - xy$ **(v)** $(x + 4y)(x - 2y) + 8y^2$

  (w) $(a + 1)(b + 1) - ab$ **(x)** $(n - 1)(m - 1) - nm$

  (y) $(2p + 3)(p + 2) - 2p^2$ **(z)** $(n - 3)(n + 1) + 2n$

> **Remember:**
> $(x + 5)^2 = (x + 5)(x + 5)$

3 $ABCD$ is a rectangle.
$AB = (x + 5)$ cm.
$BC = (x + 6)$ cm.
Show that the area, in cm^2, of
$ABCD$ is $x^2 + 11x + 30$.

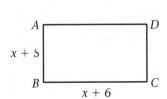

4 *PQRS* is a rectangle.
$PQ = (x + 3)$ cm.
QR is 5 cm shorter than *PQ*.

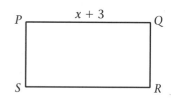

(a) Show that the area, in cm², of *PQRS* is given by the expression $x^2 + x - 6$.

(b) Work out the area of *PQRS* when $x = 8$.

Exercise 28.3   **Links 28D, E**

You will need graph paper and tracing paper.

1 (a) Copy and complete the table of values for $y = x^2$.

(b) Draw the graph of $y = x^2$ for values of x from -3 to 3. Label this graph **A**.

x	-3	-2	-1	0	1	2	3
y		4		0	1		9

(c) On the same axes, draw and label the graphs
B $y = x^2 + 4$ **C** $y = x^2 - 3$ **D** $y = 2x^2$

2 (a) Use tracing paper to make a copy of this graph of $y = x^2$ for values of x from 0 to 4.

(b) Complete the graph for values of x from 0 to -4.

(c) On the same axes, draw and label the graphs

B $y = -x^2$

C $y = -x^2 + 3$

D $y = -\dfrac{x^2}{2}$

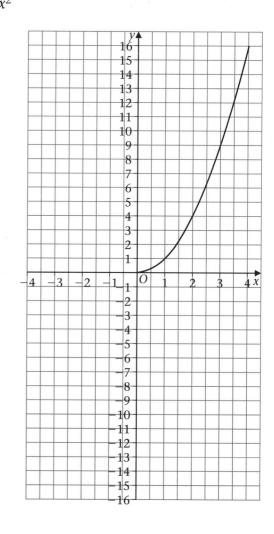

3 Here are six graphs labelled **A**, **B**, **C**, **D**, **E** and **F**.

A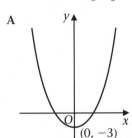
(0, −3)

B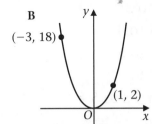
(−3, 18)
(1, 2)

C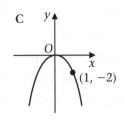
O
(1, −2)

D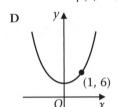
(1, 6)

E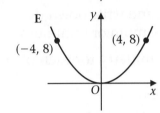
(−4, 8)
(4, 8)

F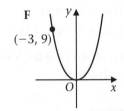
(−3, 9)

Here are six functions labelled ①, ②, ③, ④, ⑤, ⑥.

① $y = x^2$ ② $y = x^2 + 5$ ③ $y = x^2 - 3$

④ $y = \dfrac{x^2}{2}$ ⑤ $y = 2x^2$ ⑥ $y = -2x^2$

Match each function to the correct sketch.

4 Draw the graph of $y = 2x^2 - 4$ for values of x from -3 to 3.

Exercise 28.4   Link 28F

The axis of symmetry is the mirror line.

1 (a) Copy and complete the table of values for
$y = x^2 - 4x + 5$.

x	−1	0	1	2	3	4	5
y		5		1			10

(b) Draw the graph of $y = x^2 - 4x + 5$ for values of x from -1 to 5.

(c) Draw the axis of symmetry for this graph.

(d) Write down the equation of this axis of symmetry.

2 (a) Copy and complete the table of values for $y = 12 - x^2$.

x	−3	−2	−1	0	1	2	3
y		8			11		3

(b) Draw the graph of $y = 12 - x^2$ for values of x from -3 to 3.

(c) Draw the axis of symmetry for the graph.

(d) Write down the equation of the axis of symmetry.

(e) Find the maximum value of y.

3 Draw the graph of each of the following.

(a) $y = x^2 - 3$ for values of x from -3 to 3

(b) $y = 15 - x^2$ for values of x from -4 to 4

(c) $y = x^2 - 4x$ for values of x from -1 to 5

(d) $y = x^2 + 6x$ for values of x from -7 to 1

(e) $y = x^2 - 2x - 3$ for values of x from -3 to 5

(f) $y = 2x^2 + 4x - 1$ for values of x from -4 to 2

In each case, draw the axis of symmetry, and write down the coordinates of the point where y is a maximum or minimum.

4 Work out the equation of the axis of symmetry for each of these quadratic graphs.

(a) $y = x^2 + 7$ (b) $y = -2x^2$

(c) $y = x^2 - 4x + 3$ (d) $y = 3x^2 - 12x + 8$

(e) $y = -x^2 + 5x + 3$

5 (a) Draw the graph of $y = 2x^2 - 8x + 5$ for values of x from -1 to 5.

(b) Find the minimum value of y and the value of x when this minimum occurs.

(c) Find the values of x when $y = 0$.

(d) Find the values of x when $y = 10$.

6 Given that $y = -x^2 + 4x + 3$, find the maximum value of y.

Exercise 28.5   Link 28G

1 Draw the graph of $y = x^2 - 5x + 4$ for values of x from -1 to $+6$.

(a) Use your graph to solve the quadratic equations

(i) $x^2 - 5x + 4 = 0$ (ii) $x^2 - 5x + 4 = 2$

(iii) $x^2 - 5x + 4 = -1$

> **Remember:**
> $x^2 - 5x + 4 = 0$ is when $y = 0$ on the graph.

(b) Write down the minimum value y can take.

(c) Write down the value of x that gives the minimum value of y.

2 Draw the graph of $y = 2x^2 - 5x - 3$ for values of x from -2 to $+5$.

(a) Use your graph to solve the quadratic equations

(i) $2x^2 - 5x - 3 = 0$ (ii) $2x^2 - 5x - 3 = 10$

(iii) $2x^2 - 5x - 3 = -3$

(b) Estimate the coordinates of the minimum point for the function $2x^2 - 5x - 3$.

(c) Give a value of y for which the equation $y = 2x^2 - 5x - 3$ does not have a solution.

3 Draw the graph of $y = 12 + 5x - 3x^2$ for values of x from -3 to $+4$.

 (a) Use your graph to solve the quadratic equations
  (i) $12 + 5x - 3x^2 = 0$
  (ii) $12 + 5x - 3x^2 = -5$
  (iii) $12 + 5x - 3x^2 = 10$

 (b) Estimate the maximum value for $f(x) = 12 + 5x - 3x^2$.

4 (a) Copy and complete the table of values for $y = 2x^2 - 4x + 7$.

x	-2	-1	0	1	2	3	4
y		13					23

 (b) Use the graph to solve each of the following equations.
  (i) $2x^2 - 4x + 7 = 13$
  (ii) $2x^2 - 4x + 7 = 16$

 (c) State clearly why there are **no solutions** to the equation $2x^2 - 4x + 7 = 0$.

 (d) Find
  (i) the minimum value of y
  (ii) the corresponding value of x.

Exercise 28.6   Links 28H, I

1 The graph shows information about the cost, weight and megapixel rating of four digital cameras.

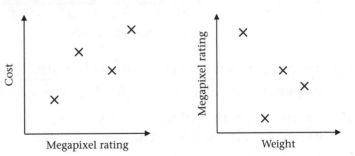

Camera A costs the most.
Camera D has the lowest megapixel rating.
Camera B costs more than camera C.
Copy and complete the graphs by identifying which cameras are represented by the points.

Identify A from the cost axis. This will tell you something about the megapixel value.

2 The graphs show information about the heights, weights, shoe sizes and IQ of four people.

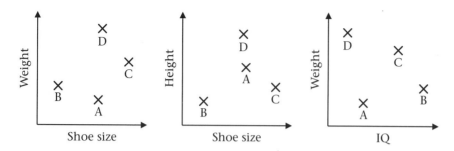

Using the evidence from the three graphs shown, construct graphs for

(a) shoe size and IQ

(b) IQ and height

(c) height and weight.

3

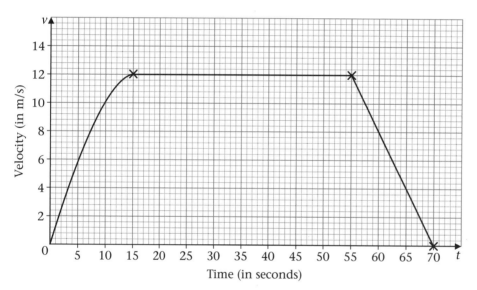

A motor cycle travels between two sets of traffic lights.

The diagram is the velocity–time graph of the motor cycle.

(a) Use the graph to find the velocity of the motor cycle 30 seconds after it leaves the first set of traffic lights.

(b) Describe fully the journey of the motor cycle between the two sets of traffic lights.

4   The graph shows the number of people inside
a football ground on match day.

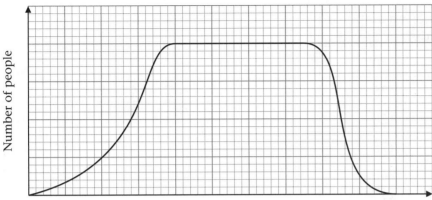

A game of football takes about 2 hours including
half time and time added on for stoppages.
The match kicked off at 3:00 pm.
Estimate from the graph the time the first person
arrived and the time when all the spectators had
left the ground.

5   Katrina boiled an egg. The sketch graph below is a graph of
the temperature of the water against time during the boiling
process.

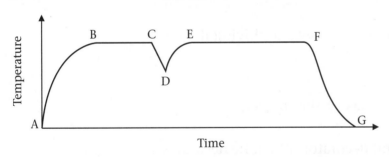

(a)   Describe fully how the temperature of the water
changed between the points marked A and B on
the graph.

(b)   What do you think happened at the time corresponding
to point C on the graph?

(c)   Describe fully how the temperature of the water changed
between the points E, F and G on the graph, stating
clearly the real-life situation you believe is represented by
this portion of the graph.

6 The table below shows information connecting the speed of a car to the shortest distance in which it can stop.

Speed (mph)	Shortest stopping distance (feet)
20	40
30	75
40	120
50	175
60	240
70	315

(a) Plot the graph of shortest stopping distance against speed.

(b) Use your graph to estimate
  (i) the shortest stopping distance for a car travelling at 45 mph
  (ii) the speed of a car when its shortest stopping distance is 300 feet. [E]

Exercise 28.7

Link 28J

1 The cost, £C, of hiring a car for a day and driving it x miles is given by the formula

$$C = 50 + 0.2x$$

Rearrange this formula to make x the subject.

2 Lee works as a painter and decorator. When he decorates a room he charges customers £15 per hour plus the cost of materials.

(a) Work out how much Lee will charge when he works for 10 hours and the cost of the materials is £80.

Val, Lee's partner, designs a formula for the charge, £C, for a job which will take h hours and for which the cost of the materials is £M.

(b) Write down what Val's formula should be.

(c) Rearrange this formula to make h the subject.

3 Rearrange each of these formulae to make the letter in brackets the subject.

(a) $y = 3x + 5$ $[x]$ (b) $p = 2q - 3$ $[q]$

(c) $a = \frac{1}{2}b + 1$ $[b]$ (d) $y = \frac{1}{4}x - 5$ $[x]$

(e) $y = ax + b$ $[x]$ (f) $v = at + u$ $[a]$

(g) $y = a + bx$ $[x]$ (h) $v = u + at$ $[t]$

(i) $C = \pi d$ $[d]$ (j) $D = 2\pi r + a$ $[r]$

(k) $b = 3(a - 4)$ $[a]$ (l) $P = 2(a + b)$ $[a]$

(m) $y = \dfrac{x + 3}{4}$ $[x]$ (n) $y = \dfrac{x - 5}{2}$ $[x]$

(o) $m = \dfrac{5n + 3}{7}$ $[n]$ (p) $m = \dfrac{3 - 2n}{4}$ $[n]$

Exercise 28.8   Links 28K, L

1 Rearrange each of these formulae to make x^2 the subject.

(a) $y = x^2$ (b) $y = x^2 + 1$ (c) $y = x^2 - 3$

(d) $y = x^2 + m$ (e) $y = x^2 - p$ (f) $y = 2x^2$

(g) $y = 5x^2$ (h) $y = \frac{1}{2}x^2$ (i) $y = ax^2$

(j) $y = bx^2 + 1$ (k) $y = cx^2 - 3$ (l) $y = ax^2 + b$

(m) $y = \dfrac{x^2}{3}$ (n) $y = \dfrac{x^2}{4}$ (o) $y = \dfrac{x^2}{a}$

(p) $y = \dfrac{x^2}{3} + 2$ (q) $y = \dfrac{x^2}{4} - 5$ (r) $y = \dfrac{x^2}{a} - 1$

(s) $y = \dfrac{x^2}{a} - b$ (t) $y = \dfrac{ax^2}{c} + b$ (u) $y = \dfrac{ax^2 - b}{c}$

2 The formula for the volume, V, of a cylinder with base radius r and height h is

$$V = \pi r^2 h$$

Rearrange this formula to make

(a) h the subject   (b) r the subject.

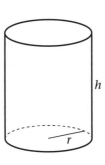

Solve questions **3** to **10** to find the value of x.

3 $x^2 + 3 = 12$ **4** $\dfrac{16}{x} - 8$

5 $2x^2 - 9 = 41$ **6** $74 = \dfrac{3}{x}$

7 $5x^2 - 27 = 153$

8 $248 - 3x^2 = 5$

9 $15 = \dfrac{120}{x}$

10 $\dfrac{30}{x} - 3 = 12$

11 (a) Show that the equation $x^3 - 5x = 32$ has a solution between $x = 3$ and $x = 4$.

(b) Use a method of trial and improvement to find this solution correct to one decimal place.

12 (a) Show that the equation $x^3 + 3x^2 = 30$ has a solution between $x = 2$ and $x = 3$.

(b) Use a method of trial and improvement to find this solution correct to two places of decimals.

13 Use a method of trial and improvement to find, correct to one decimal place, the positive solution of $x^3 + 3x = 40$.